The Digital Connection

The Digital Connection

A Layman's Guide to the Information Age

Irwin Lebow

COMPUTER SCIENCE PRESS
An Imprint of W. H. Freeman and Company / New York

Library of Congress Cataloging-in-Publication Data

Lebow, Irwin.
The digital connection: a layman's guide to the information age /
Irwin Lebow.
p. cm.
ISBN 0-7167-8203-0
1. Digital electronics—Popular works. I. Title.
TK7868.D5L398 1991 89-23438
621.382—dc20 CIP

Printed in the United States of America

Computer Science Press
An imprint of W. H. Freeman and Company
41 Madison Avenue, New York, NY 10010
20 Beaumont Street, Oxford OX1 2NQ, England

1 2 3 4 5 6 7 8 9 0 RRD 9 9 8 7 6 5 4 3 2 1

To Grace

Her mouth is full of wisdom,
Her tongue with kindly teaching.

Proverbs 31:26

Contents

2 *Telecommunications* 33

3 *Sound, Music And Speech* 58

Preface

Just as past ages were called "bronze" or "iron" after the dominant technology of the times, so our age is often characterized as the "information age." For unquestionably the current computer and telecommunications "information explosions" have dominated our technology and thereby made irreversible changes in the way we live and work. Yet aspects of this technology are mysterious to most of us. Why has the computer become so ubiquitous? What is the basis for the improvements in the quality and convenience of our telephone service? How do automatic teller machines work? Why do those marvelous compact discs sound so clear?

At the core of this vast information explosion is its "digital connection," the common dependence of modern computing, telecommunications, and recording upon digitization, the technique of representing information by numbers. Once in digital form, the information can be processed in a computer, transported from one place to another and recorded and played back, all with extremely high accuracy. Without some understanding of its digital underpinnings, one cannot begin to appreciate the ramifications of the information revolution or the implications of the emerging information society.

The idea of writing a "layman's guide to the information age" came after spending many hours explaining these topics to my colleagues in industry, individuals with expertise in a wide range of disciplines, but with little or no technical training. This experience convinced me that many people without technical training were eager to understand what's behind information technology. I was finally impelled to write this book when I discovered that, despite the plethora of books on computers, none of them included modern telecommunications and recording, all from the vantage point of their common digital roots.

Throughout the book, I have emphasized these common roots and the resulting mutual dependence of communicating, computing, and recording. Thus, the discussion on computers shows not only how they work but also why they are so dependent on telecommunications. Similarly, the book not only explains why Shannon's information theory is so fundamental to communication, but also shows how putting this theory into practice was made possible by the burgeoning digital computer technology. And the discussion on the compact disc and its technical and commercial successes shows the crucial role played by both computing and telecommunications technology.

I have tried to address these topics in an informative and entertaining way, using intuition and illustrations to support the logical arguments in place of the customary equations and technical jargon. Naturally, intuition depends upon experience. The more familiar someone is with a subject, the better his or her intuitive feeling for the subject will be. For this reason, the reader with less background may want to skim the more technical sections on first reading and return to them later.

Many years ago, C. P. Snow pointed out the imbalance in our culture—the fact that those in the sciences had available to them the culture in both the scientific and nonscientific worlds, while the scientific domain was all but closed to the non card-carrying scientist. Like it or not, science and technology are increasingly important parts not only of our economy but of our culture as well. I hope that this book will enable you to leap across the barrier and into the heretofore forbidden world of information technology.

Irwin Lebow
July, 1990

Acknowledgments

There are many who provided advice, assistance and encouragement while I was writing this book and preparing it for publication. William Lebow and David Lebow read portions of the text and made helpful comments. William Lebow also performed the calculations needed to demonstrate the effect of harmonics in reproducing musical sounds. Vinton Cerf read some early material and made some helpful comments. Alan McLaughlin, Clifford Weinstein and Michael Hart, former colleagues at the MIT Lincoln Laboratory, were kind enough to dig into their archives for photographs that I hope have added interest to the text.

Andrew Oram read the entire first draft manuscript critically and offered many very useful comments, criticisms, and suggestions. A talented technical writer with a background in music, Andy was my model of the intelligent layman to whom the book is addressed. If you find the book helpful, it is due, in no small way, to his insightful comments.

I am also grateful to the people at Computer Science Press and W. H. Freeman for their assistance and cooperation. Special thanks are due to Barbara Friedman in the early stages and later, to William Gruener. I am especially grateful to my editor, Diane Maass, who worked so hard in helping get the book in shape for publication. We may not have always agreed on everything, but I am sure that the book is the better for the many interactions I had with Bill and Diane.

Writing a book represents a large commitment of time and both physical and psychic energy. While I had considerable motivation on my own, it was my wife, Grace, who encouraged me to undertake it and then kept me at it even when my spirit waned. This constitutes only the most recent of the many ways in which she has always been at my side for so many years.

The Digital Connection

1

Digital Computing

The digital computer is the very symbol of the information age. Yet, digital computing is as old as the first man who learned to add and subtract with the aid of the digits on his hands and feet. The modern digital computer uses these same basic principles. It represents all quantities as digits or whole numbers and performs calculations by manipulating those digits much as we do when we compute by hand.

In contrast, an analog computer represents all quantities by proportional physical measurements, such as distances along a line or electrical voltages, and performs calculations by manipulating those measurements. The slide rule is an example of a simple form of analog computer because it uses distances along a scale to represent the numbers to be manipulated.

An analog computer, however, always has some practical accuracy limits. The accuracy of a slide rule, for example, is about one part in a thousand and is limited by how accurately a person can read the engraved scale. While most electronic analog computers are considerably more accurate than manual slide rules, each improvement in accuracy is successively more difficult to achieve. Digital computers have the fundamental advantage over analog computers in their ability to perform calculations with arbitrarily high accuracy. You know that, in principle, you can multiply 20-digit numbers as easily as 3-digit numbers; it's just more tedious and takes longer. The same thing holds true in a digital computer, only the computer doesn't mind the tedium.

The Evolution of the Digital Computer

The modern digital computer resulted from the efforts of several groups of pioneers during the 1930s and 1940s. The basic idea of the automatic digital computer is due to Charles Babbage, an English mathematician who lived in the first part of the nineteenth century. His machines, called the *Analytical Engine* and *Difference Engine*, were mechanical. The most famous early electronic digital machine, widely considered to be the true forerunner of the modern computer, was the *ENIAC*, constructed in 1946 at the Moore School of the University of Pennsylvania by a team led by John Mauchly and J. Presper Eckert. It is now generally agreed that the fundamental ideas underlying the ENIAC were first demonstrated in the work of John V. Atanasoff of Iowa State University between 1937 and 1942. The basic structure of the computer as we know it today was developed by the late mathematician, John von Neumann, of the Institute for Advanced Study in Princeton. Almost all modern computers use this structure and are referred to appropriately as *von Neumann machines*. It is interesting to note in passing that Von Neumann was one of an unusually large number of outstanding scientists who came to the United States from Nazi-dominated Hungary just before World War II. I recall as a student in the late 1940s how physicists viewed this "Hungarian phenomenon" as akin to a violation of the second law of thermodynamics. While nuclear physics and atomic energy have been the best publicized of the areas in which these emigre scientists contributed to the United States, Von Neumann's contribution to the embryonic digital computer field was no less significant.

The analog computer was well established by the time of these earliest digital computers. But it did not take very many years before the digital computer began to supplant the analog computer in many application areas. Of course, one of the reasons was the superior accuracy of the digital machine. But perhaps even more important was the astonishing rate at which advances in electronic component technology increased the power and decreased the cost of digital computing. This technology included the development of first the transistor, and later the *integrated circuit*, a small wafer or chip of silicon on which large numbers of transistor circuits are fabricated. The analog computer also benefited from these component advances, but to a much lesser extent.

The first integrated circuits were fabricated in the 1960s. Progress was rapid as the numbers of transistors on a chip increased from tens to hundreds to thousands and now millions. By the 1970s, they were

Figure 1-1 The Eclipse of the Slide Rule (From the Washington Post Writers Group)

sufficiently advanced to permit a pocket-size calculator to be built at a reasonable cost, followed shortly thereafter by the earliest desktop computers. The analog computer has all but disappeared. Even the slide rule, long the workhorse portable computing aid, had to give way to the pocket calculator, the phenomenon that the Bloom County comic strip captures so cleverly in Figure 1-1.

Representing Analog Measurements

Some kinds of data fit naturally into the digital or whole number representation fundamental to the digital computer. Monetary values are a good example of this category because dollars and cents are measured in discrete units. Prices do not vary over a continuum of values, but in discrete steps.

But how about when we are dealing with analog data—that is, data that can take on a continuum of values? How can these data be represented digitally? Let's look at an example to illustrate how this is done.

Suppose you have the need to measure the dimensions of your living room for a wall-to-wall carpet installation. Once you have made the measurement, you can use a pocket calculator to compute how much carpet is needed and what it will cost. If you have large enough feet, you could walk the length of the room heel to toe and count the number of "feet," making the assumption that your foot is roughly 12 inches long. Another, more accurate technique would be to use a steel tape. In the first case, the accuracy is perhaps plus or minus 3 inches. In the second case, the accuracy is about plus or minus 1/8 inch, about 25 times more accurate than the first. Suppose that the first measurement technique estimated the length as 15 feet 2 inches or, in decimal notation, 15.16667 feet. The steel tape measurement estimated the length as 14 feet 11-7/8 inches or, in decimal notation, 14.9895833 feet.

Any representation of a physical measurement by a number makes a statement about the accuracy with which you know the parameter being measured. Most of the figures to the right of the decimal point in our example measurements are an illusion of accuracy. Since the first measurement was accurate only to within 3 inches, any digit beyond the first place to the right of the decimal point is meaningless; we should call the measurement 15.2 feet. In the second case only two decimal places are warranted, giving us 15.00 feet. Since we will ultimately use this number in a digital computer, we would use three decimal digits for the crude measurement and four for the more accurate measurement.

Thus, even though the length of the living room can have a continuum of values, the number of possibilities is limited by the accuracy with which the quantity is measured. The accuracy must be sufficient for whatever computation is to be made.

Representing the length of a room with three decimal digits is equivalent to saying that the length can have one of the 1000 possible values from 00.0 feet to 99.9 feet. When we use four decimal places, there are 10,000 possible values from 00.00 to 99.99 feet. In this way, data with a continuous range of possible values are represented by a discrete range of numbers. The general rule is this: 1 decimal digit can specify one of 10 things; two digits can specify one of 100 things; three digits can specify one of 1000 things etc. It doesn't matter what the things are.

The Pocket Calculator

Despite its apparent complexity, the computer is remarkably simple conceptually. Many things in science and technology are like this. Often the mathematics involved will frighten the uninitiated, but not so with computers. No knowledge of advanced mathematics is needed to understand how a computer works. So let's begin with the simplest computer, the pocket calculator.

Any pocket calculator can perform the elementary operations of addition, subtraction, multiplication, and division; every computer has some form of calculator. For example, to add two numbers, say 15 and 17, you punch in

$$15 + 17 =$$

and the calculator displays the result, 32, almost instantaneously.

The calculator does its arithmetic just the way we were taught to do it in school. To add two numbers, we write the numbers down one above the other

$$\begin{array}{r} 15 \\ \underline{17} \\ 32 \end{array}$$

and perform the addition using an addition table that we've memorized in our earliest years. The pocket calculator does exactly the same thing. The two numbers are entered into the calculator's *registers,* where a register is a device that can *store* a number. The calculator uses a stored addition table to perform the addition. The sum is also stored in a register.

Information Storage

This concept of information storage is fundamental. The ancient Sumerians kept records by chiseling wedges into stone tablets. The discovery in ancient Egypt that markings could be made with ink on papyrus constituted a breakthrough in storage technology. Everyone is familiar with mechanical or electrical devices that are used for storage. An automobile odometer is a digital storage device; it stores a number by the positions of a series of wheels on which are engraved the digits 0-9. A phonograph record is an analog storage medium because it stores a replica of an audio signal in a groove cut into its surface. A magnetic tape can be either an analog or a digital storage medium since it can store either a replica of an audio signal or digital numbers that represent the signal.

There has been a tendency to ascribe human-like characteristics to computers. Almost from the start, the ability of a computer to store things has been referred to as *remembering*, and a set of storage registers is almost always called a *memory*. Call it what you will, storage or memory is an essential part of the computer. An important reason for the speed with which the digital computer replaced the analog computer is because numbers can be stored more rapidly, efficiently, and conveniently in digital form than in analog form.

An Example: Calculation

Let's look at a typical calculation that might be performed with the aid of a pocket calculator. Suppose you own the portfolio of stocks shown in Table 1-1 and would like to compute its value. To obtain the result, you have to multiply the number of shares of each stock by its price and then take the sum. Since there are six stocks in your portfolio, this

Table 1-1 A Stock Portfolio

Company	*Number of Shares*	*Price per Share*	*Value*
AT&T	1000	$ 24 5⁄8	$ 24625
Digital	500	167 1⁄8	83562.50
Ford	300	100	30000
General Motors	600	89 3⁄4	53850
IBM	100	163 3⁄4	16375
PEPSICOLA	300	32 1⁄4	9675
			$218,087.50

requires six multiplications followed by the addition of the resulting six numbers. You might do it in the following way:

1. Multiply the number of shares of AT&T by the price per share.
2. Write down the result.
3. Multiply the number of shares of Digital by the price per share.
4. Add this result to the result in step 2.
5. Write it down.

Repeat steps 3–5 for Ford, General Motors, and IBM.

• • •

15. Multiply the number of shares of PEPSICOLA by the price per share.
16. Add the result to the result in step 14.
17. Write down the final result.

These 17 steps constitute a *program* for the desired calculation. But the program, as stated, glosses over the precise way in which the pocket calculator must be used. In reality, each of the steps involving an arithmetic operation consists of three steps: (1) enter the first number, (2) enter the operation (+ or ×), followed by the second number, and (3) press = to perform the calculation. You would probably do it much the same way by hand as you would with a calculator. So why use a calculator? The first reason is accuracy: if you had to do all of the additions and multiplications by hand, chances are you would make a mistake. The second reason is speed: As soon as you press the = button on the calculator, your result immediately appears in the calculator's display. When you use a pocket calculator, you spend most of your time punching in the numbers and writing down the results. When you do the calculations by hand, you spend most of your time performing the multiplications and additions.

But even when we use a calculator, the process is tedious, error-prone and, because we still have to enter each number and operation separately. How can we speed up this process? The calculator performs its tasks very rapidly. But these tasks are limited to the arithmetic operations. It will not help to speed up the arithmetic—it hardly matters whether it takes a calculator a millionth or a billionth of a second to add a pair of numbers. What we need is some way of speeding up all of the slow manual operations sandwiched between the calculations.

The Stored-Program Computer

One way to speed up the portfolio calculation shown on the previous page would be to store within the computer the equivalent of the 17 program steps. Then the sequence of instructions could be performed very rapidly without the delays inherent in the manual operations. This is precisely what a digital computer does. It is, in fact, called a *stored-program* computer precisely because it can store a program, i.e., a sequence of *instructions,* the equivalent of the steps in the example. The great contribution of the pioneering efforts of the 1940s was the development of this stored-program structure.

It is easy to see how a computer might store a number, but how it might store an instruction is less obvious. When we store something on paper, we can write down letters, words, numbers, or anything else we would like, so long as we use a *language* that is generally understood. We can in this way direct a person to perform some task by writing the necessary instructions down on paper.

The computer needs the same capabilities. But the storage mechanism in a computer, the register, stores only numbers. To permit a register to store something other than numbers, we need a language in which numbers are associated with the things to be stored. For example, all the characters on a typewriter keyboard can be stored in computer registers by assigning a unique number to each character. The computer "knows" the number representations and can therefore interpret them as keyboard characters. A number can be assigned to represent each instruction type. The computer then stores those numbers to represent the instructions. With the knowledge of this *instruction language* the computer can interpret the instructions. For instance, suppose that a computer can execute 100 different instruction types. By assigning each a unique number between 00 and 99, the computer can then represent each type uniquely with a two-digit number. For example, the number 01 might designate an addition and the number 02, a multiplication.

Several sets of storage registers are required to perform the portfolio calculation with a calculator, which in computer terminology is usually called an *arithmetic-logic unit*. The data are stored in two banks of registers called *tables*, one for the stock prices and the other for the holdings. The instructions are stored in another table.

With this structure in mind we can then rephrase the pocket calculator steps in the form of instructions to the computer based upon the way the computer stores information. The first instruction then does the following:

> *Fetch the number of shares of AT&T from the holdings table; fetch the AT&T share price from the price table; multiply the two numbers.*

Similarly, the second instruction does the following:

> *Store the result of following Instruction 1 in a temporary storage register.*

The third instruction:

> *Fetch the number of shares of Digital from the holdings table; fetch the Digital price per share from the price table; multiply the two numbers.*

The fourth instruction:

> *Add the result to the number stored in the temporary storage register.*

The fifth instruction:

> *Store the result of Instruction 4 in the temporary storage register.*

The remaining instructions continue in the same way. At the end of Instruction 17, the desired result appears in the temporary storage register and can be displayed to the user.

The key to this process is the organization of the computer in which the instructions that make up the program are stored in one set of registers and the data stored in another set. This permits the program to run very rapidly. Of course the instructions and the data must be inserted into the registers to begin with, and this involves manual operations. If you wanted to perform this portfolio calculation only once, then there would be little point to use a computer. But if you wanted to perform the calculation every week to reflect changing stock prices, you would need only to insert the new prices, press a button, and a new portfolio value would appear very quickly.

Binary Numbers

Any storage medium makes use of some physical process. Magnetic tape stores a digit by magnetizing a spot on the tape in a direction determined by the digit being stored. If this spot were storing a decimal digit, 10 directions and/or intensities of magnetization would have to be defined. A more reliable way to store digits on magnetic tape is to define two directions of magnetization rather than 10. It is similarly easier to build an electronic circuit that has only two stable states than it is to build one with ten stable states. Because of this, all basic storage elements within a computer are built to represent one of two numbers, rather than one of 10. This property of storage elements has led to the use of the binary rather than the decimal number system to perform arithmetic operations. The following limerick says it all:

An inventive young man named Shapiro,
Became a sensational hero,
By doing his sums,
Using only his thumbs,
And calling them One and Zero.

Table 1-2 Binary and Decimal Equivalents

Binary	*Decimal*
0	0
1	1
10	2
11	3
100	4
101	5
110	6
111	7
1000	8
1001	9
1010	10
1011	11

We might well have developed a binary system for our daily business had evolution left us with only a thumb on each hand. With only two digits instead of 10, it would have been only natural to compute with the two digits 0 and 1, rather than with the ten decimal digits 0 through 9. Table 1-2 shows the binary equivalents of the first few integers.

The binary digit, abbreviated *bit*, is fundamental to the way that computers work. By analogy with decimal numbers, one bit can represent one of two things; two bits, one of four things; three bits, one of eight things, etc. For every bit added, the number of items that can be represented is doubled, as shown in Table 1-3. For example, four binary digits are required to represent the fingers of your two hands. The number of bits required to represent a physical measurement is determined by the accuracy of the measurement: three decimal places require 10 bits (1024), four decimal places require 14 bits.

A group of eight bits is called a *byte*. A typical register length in a small computer is 16 bits or 2 bytes, and it can hold 256×256, or 65,536 different numbers. Larger computers generally use 32-bit registers that can store more than 4 billion numbers.

Table 1-3 Representation by Binary Digits

Number of Bits	*Number of Items*
1	1
2	4
3	8
4	16
5	32
6	64
7	128
8	256
9	512
10	1024

How a Computer Does Arithmetic

The arithmetic-logic unit performs arithmetic in a computer. I noted earlier that a calculator performs arithmetic just the way you and I do. It is illustrative to demonstrate why this is so.

Let's take an addition example: the computation of the sum of the two numbers 3875 and 2462. We add the numbers digit by digit, beginning at the right-hand column. For each column, we compute a *sum* and a *carry* into the next column to the left. We continue adding numbers until we reach the leftmost digits. The calculation is shown below as a two-step process: first the computation of the carry into the column to the left; and then, the computation of the sum of the three digits in that column.

	3 8 7 5
	2 4 6 2
Carry in	1 1 0
Sum	6 3 3 7

We know what to do because we have learned the rules of addition and the addition table.

Sum

	0	1	2	3	4	5	6	**7**	8	9
0	0	1	2	3	4	5	6	7	8	9
1	1	2	3	4	5	6	7	8	9	0
2	2	3	4	5	6	7	8	9	0	1
3	3	4	5	6	7	8	9	0	1	2
4	4	5	6	7	8	9	0	1	2	3
5	5	6	7	8	9	0	1	**2**	3	4
6	6	7	8	9	0	1	2	3	4	5
7	7	8	9	0	1	2	3	4	5	6
8	8	9	0	1	2	3	4	5	6	7
9	9	0	1	2	3	4	5	6	7	8

Carry

	0	1	2	3	4	5	6	**7**	8	9
0	0	0	0	0	0	0	0	0	0	0
1	0	0	0	0	0	0	0	0	0	1
2	0	0	0	0	0	0	0	0	1	1
3	0	0	0	0	0	0	0	1	1	1
4	0	0	0	0	0	0	1	1	1	1
5	0	0	0	0	0	1	1	**1**	1	1
6	0	0	0	0	1	1	1	1	1	1
7	0	0	0	1	1	1	1	1	1	1
8	0	0	1	1	1	1	1	1	1	1
9	0	1	1	1	1	1	1	1	1	1

The addition table contains two parts, one for the sum and one for the carry. Across the top and down the left side of each table are the digits to be added. The entries in the tables give the sum and the carry digits, respectively. In the illustration, shown in boldface, 7 and 5 sum to 12; the sum table shows the entry 2 and the carry table shows the entry 1.

The computer performs addition in exactly the same way. It takes the numbers to be added digit by digit starting at the right end. For each digit, it computes the sum of the two digits plus any carry in from the right, and it generates a carry to the left using addition tables like those shown above. It mechanizes the processes by using electronic elements that can sense the digits to be added and then enter those digits in tables to extract the results.

There is only one difference. While we use the decimal system, computers use the binary system, in which the only allowable digits are 0 and 1. The decimal addition tables shown above compress to the following very simple binary tables:

Sum

	0	1
0	0	1
1	1	0

Carry

	0	1
0	0	0
1	0	1

The calculator must determine whether the digits to be added are 0s or 1s, and these, together with a carry into the digit determines the sum and the carry out. In the early days of computers, each individual storage element and each sensing element was implemented in a separate electronic component. This made the arithmetic-logic unit large and expensive. Over time, the solid-state technology advanced to the point that an entire arithmetic-logic unit could be implemented on a single chip of silicon, drastically reducing both the size and the cost of the unit. This, together with equivalent advances in the fabrication of memory elements led to the development of the *microcomputer*, which, in turn, has paved the way for the desktop or personal computer.

Computer Memory

The computer in an electronic wristwatch has only a few registers. The computer that regulates a microwave oven or the fuel flow in an automobile engine might have a few hundred or a few thousand registers. Counting the registers in the personal computer in your office or home, is more complex because there are different kinds of memory for different purposes.

An arithmetic-logic unit itself contains a few storage registers to hold numbers while a computation is being carried out. These registers are the most complex of all: they not only store numbers, but also manipulate them in the performance of arithmetic operations. For this reason they must operate at the highest possible speeds. A computer contains a relatively small number of these elements, and they are not usually called memory.

The term *memory* designates a set of registers that do nothing but store numbers. A *memory unit* is defined as a set of such storage registers, numbering in the hundreds of thousands or even millions or billions.

All memory types are not equal. They differ not only in size, but also in speed and other operational characteristics, and in cost. These differing properties have led to several different ways to categorize memory units. One important category is the way in which a register in the memory is made available for retrieval of its current contents or for storage of new information, called the *access method*. The two major access methods are *random* and *sequential*. In a random-access memory, there is no preferential ordering: it takes the same time to access any register regardless of the location of the register accessed previously. In a sequential-access memory, it is necessary to access some or all of the registers in sequence before reaching a desired register. In this type of memory, the time it takes to access a register varies, but is substantially greater, on average, than in a random-access memory. However, sequential access memories are much cheaper. Random-access memory generally costs under 0.001 cent per bit, while sequential access costs less than 10 percent of that.

The main memory of a computer is random-access. Its registers are electronic; i.e. built out of integrated electronic components. Not too many years ago, electronic registers were far too expensive to use in quantity. The workhorse random-access memory in those days was the magnetic core memory, which was built out of an array of small, donut-shaped magnets. Some programmers still refer to the main memory as "core" even though the magnetic core memory is a thing of the past.

A common example of a sequential-access memory is magnetic tape. The tape stores registers sequentially from end to end. To access a particular register, it is necessary to move the tape sequentially through all the locations between the current position and the destination. Magnetic tape used to be the primary bulk storage medium used in most computers, but has now been replaced by the magnetic disk, except for archival storage. The disk is also a sequential-access storage medium, but has certain attributes that permit faster access time. The disk registers are located in concentric rings. The magnetic head used to access the memory can move radially to select the ring on which a particular register is located, and then the disk rotates the desired register under the head. This property makes the disk faster than tape. Nevertheless, magnetic disk sequential- access memory still is slower than the random-access electronic memory. Access time is measured in milliseconds (thousandths of a second) rather than in fractions of microseconds (millionths of a second).

In a personal computer, the disks may be hard or floppy. The two types are quite similar in principle. The thin, flexible, floppy disk is very cheap, but is limited in speed and capacity. Hard disks have greater speed and capacity and thus, permit the computer to solve more complex problems in less time.

Another way to categorize memory is by the permanence of its stored data. The *compact disc* (*CD*) is a good example of *read-only* memory. Once a recording is made, it becomes a permanent record and is never changed. In contrast, magnetic disks and tapes, which have *read-write* memory, may be erased and reused many times. Optical discs similar to compact discs are widely used as read-only memories because of their ability to store huge amounts of data in a very small space. A newer technology combining optical and magnetic phenomena permits optical discs to be used as read-write storage media. Electronic memories are usually read-write, although read-only memories are sometimes very useful. Read-write memories are more versatile, but if you want to prevent accidental erasure, it is preferable to use read-only memory.

Memory may also be classified by whether it is *volatile* or *nonvolatile*. A volatile memory "forgets" the data stored in it when the electric power is turned off. A nonvolatile memory retains its information after the power is turned off. Read-only memories of all kinds are nonvolatile by their very nature, and most sequential-access memories are nonvolatile. For example, floppy disks store data magnetically, and this magnetization remains even after the power is turned off. Random-access read-write memories being electronic lose all their information when the power is removed from the computer. When you save your data on a computer before turning the power off, you are transferring data from volatile random-access memory to nonvolatile disk memory. If you forget to do this, you lose your data when the power is turned off.

By far the fastest and most flexible kind of memory is electronic, random-access. But since it is also the most expensive, its size is limited. Because of this limitation, random-access storage is augmented with slower forms of memory. When the computer exhausts its supply of electronic memory, it may transfer a block of old information from electronic memory to disk, replacing it with a new block from the disk. This *memory swapping* is time-consuming and is minimized by making the electronic memory as large as possible.

Computer Hardware

Computers fall into two broad classes, *general-purpose* and *special-purpose*. The two classes work in essentially the same way, and both have the following constituent elements.

1. An arithmetic-logic Unit: The part of the computer in which calculations are performed.
2. Memory units: One or more banks of registers that store data or programs.
3. Input/output units: An input device permits the information to be inserted into the computer (e.g., a keyboard); an output device permits information to be extracted from the computer (e.g., a display screen or a printer).
4. A control unit: the part of the computer that orchestrates the many details necessary to ensure correct operation.

These main constituent elements all contain storage of one kind or another. The computer works by transferring information from one element to another. For example, you insert data into the computer by transferring the data into memory via an *input device* such as a keyboard. In the reverse process, data are transferred from memory to an *output device* such as a display monitor or a printer. Arithmetic is performed by sending the data from memory to the *arithmetic-logic unit*. The results of the calculation are then returned to memory. Instructions in a program are interpreted by the *control unit*, which, in turn, directs events to occur in the right sequence by sending commands throughout the computer at extremely high speeds.

A special-purpose computer has only a single function. Its memory contains a program that permits the computer to do a single calculation, or a few at most. Its memory and input/output units are tailored to this function. For example, the computer in a microwave oven keeps time and sets the intensity of the microwave radiation. The computer under the hood of an automobile adjusts the mixture of gasoline and air.

Although these special-purpose computers based upon the microcomputer are virtually everywhere, they are unobtrusive. General-purpose computers are much more in evidence. Since a general-purpose computer must be able to do a large variety of different tasks, its component elements must be configured for this purpose. In terms of this structure, these component elements must be able to perform a large number of different instructions. In particular, the control unit must be able to command the machine to perform each instruction in whatever sequence the program directs. In the microprocessor, the heart of the personal computer, the control and arithmetic-logic units are combined into a single large-scale-integrated circuit called the *central processing unit*, abbreviated *CPU*, but often referred to as a "computer on a chip." The designations of these CPU chips describe the computer's power. Thus, the 286 machine and 386 machine, two of the more common personal computer classes, are named after the Intel 80286 and 80386 chips.

Special-purpose computers have the same general-purpose structure, but with fixed programs. They use the same microprocessors as do the general-purpose computers because of their mass-production based economy, even though their generality may not be needed. Thus both special- and general-purpose computers have benefited from the enormous advances in technology over the last 40 years. The same technology that brings a powerful personal computer into your office or home for a couple of thousand dollars has reduced the cost of a compact-disc player to under $200. This technology includes faster arithmetic circuits and larger, cheaper and more versatile memories.

All of this remarkable technology has brought the computer into virtually every part of our society and has transformed it irreversibly, sometimes in ways that have had profound societal implications. John Naisbitt, the author of *Megatrends* was especially concerned with the societal implications of the microprocessor when he wrote: "Earlier computer technology could be applied to some products, electronics and large-scale office information equipment, but not others. Microprocessors can improve almost anything, anywhere, and are consequently far more threatening [to the established order]."

Computer Software

Hardware alone is not enough. Any computer, large or small, must be programmed to perform its tasks. Special-purpose computers are programmed once to do one thing. While in theory, a general- purpose computer can perform any calculation at all, every computer has certain limitations (e.g., speed and memory capacity) that restrict its practical applicability. Within these limitations, a general-purpose machine can do many different things, as long as it stores a program telling it what to do. The computer's usefulness depends on how easy or difficult it is to write these programs.

Machine instructions permit the programmer to do arithmetic and store and retrieve data—in short, anything necessary to solve any problem. All one must do is to write the program or software—i.e., the appropriate sequence of instructions—and the computer will do the rest. It sounds easy. But programming is not easy at all. Writing a program is a logical process. The computer does not think for itself. On the contrary, it will follow the programmer's directions to the letter. If the programmer writes a correct program, the computer will produce the correct answers. If even a tiny mistake in logic is made, the computer will give incorrect answers.

If programmers had to use the primitive instructions built into the computer hardware, programming would be extraordinarily intricate and time-consuming. Instead, *programming languages* have been developed that simplify the process in many ways. In keeping with the diagram below, you can think of a programming language as being a

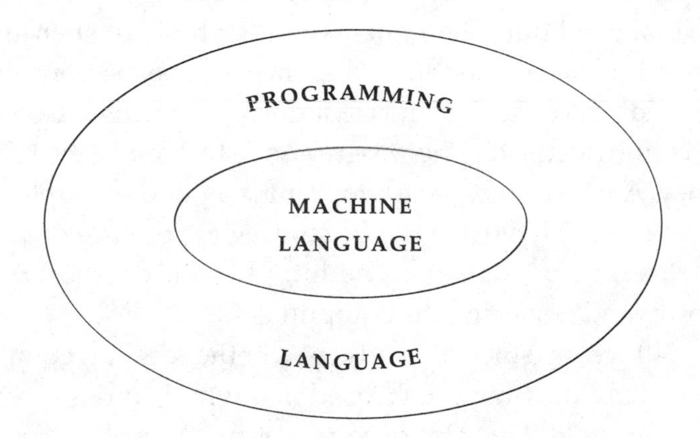

shell surrounding the machine language that encompasses the machine instructions. Usually, a programming language instruction combines several machine language instructions to perform some higher-level function. Two of the most widely used programming languages, FORTRAN and COBOL, were devised for mathematical and business programs, respectively. These and other programming languages are somewhat closer to a natural language, such as English, than is the machine language, and are therefore easier for the programmer to use.

Even with the increased sophistication of these languages, programming remains difficult, precise, and laborious. The world is filled with computer programs that *almost* work or that work *most* of the time. When considering a problem to be solved on a computer, a programmer has to think through every possible contingency. Otherwise, the computer may do something strange under certain circumstances. Some programming errors are gross. Such an error might cause the computer to access the wrong memory file and might even lead to the interpretation of data as instructions. But the very magnitude of this kind of malfunction guarantees that it would be easily detected by the programmer.

Much harder to deal with are the subtle little errors in logic that affect the program operation only a small percentage of the time under certain special conditions. It might take a long time for such an error to become evident. An error of this kind, once it appears, may also take a long time to diagnose and correct. Modern programs are so complex that it is very difficult for the programmer to handle every possible contingency. A word processing program is a good example of a very complex program. I have had the frustrating experience of having my keyboard "freeze up"; there was nothing I could do short of turning the power off and restarting the computer.

In the 40 years since the advent of the digital computer, its hardware capabilities have increased manyfold and its costs have decreased manyfold. Our ability to program computers has also improved over the years. New computer languages and programming tools help the programmer immeasurably. But these improvements in software technology are quite modest compared to the hardware advances. The biggest costs for large scale computer users today are software, because of its labor intensiveness. The technology behind the hardware advances has permitted ever more elementary circuit elements to be fabricated on a single chip of silicon. While this technology is astonishing in its technical sophistication, the most advanced computer chips are logically simple. In contrast, the writing of a computer program remains an intricate exercise in logical thought.

Most computer programmers don't have any special technical background. The fundamental reason that computers are so accessible to so many is quite simple: the ability to understand and use computers does not require any special engineering or mathematical knowledge. What it does require is the ability to think logically and clearly. Technology is limited in its ability to make the human process of logical thought any easier.

Using Computers

With the proliferation of computers, the number of programmers has increased manyfold. Even more significantly, many nonprogrammers have become computer users ever since the personal computer became cheap enough to be common in the office and later at home. The average person sitting in front of a personal computer is not a programmer, but a user. One can think of these user-oriented programs as providing an additional shell around the programming language, as shown in the diagram on the next page.

The instructions available at the user level should be easy to understand and remember. They should be independent of any of the underlying structure of either the computer hardware or the software with which the program was written. But even so, the user must follow the rules that the programmer has stipulated for operating the program. The program has no choice but to follow the instructions given to it by the user. If the user does the correct things, the correct results are obtained; if the user does something wrong, the results are unpredictable.

How hard is it to get to the point where you can use a program with some facility? That depends upon both the user and the program.

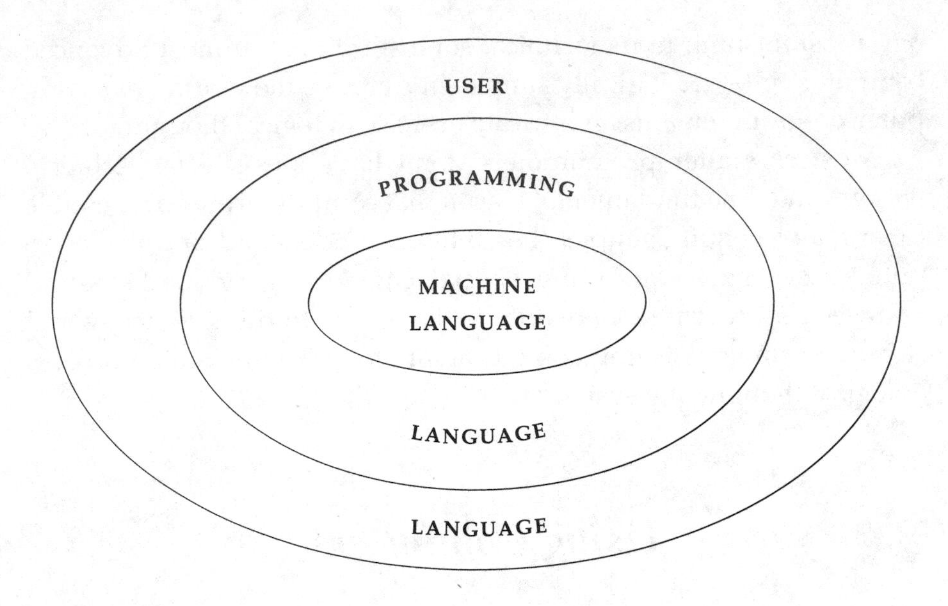

The term *user-friendly* describes a program that is relatively easy to learn. The opposite, *user-hostile,* is still the way most novice users would describe many programs, even some that call themselves user-friendly. One important feature of user-friendliness is clear, readable documentation. Another of the primary features of user-friendliness is helpful messages presented to let the user know when he has done something wrong and to give him clues as to how to get out of the predicament. I expect that any of you who have used personal computers echo the sentiments of the user in Figure 1-2 when confronted with a message that doesn't say quite what you expected.

Another problem that many people have with even friendly computer programs is communication. We are used to exchanging information with people not machines. People tell us immediately to speak more clearly if we mumble, and they can often tell what we mean even if we say it incorrectly. Unfortunately, computers are not as flexible as humans. They are completely literal. They do just what we tell them to do, not what we mean to tell them to do. We are not used to com-

Hit any key to continue

Figure 1-2 A Brief Message from a "User-Friendly" Program

municating with entities that require us to be absolutely precise and correct all the time.

A good example of a computer/communications program that most of us use more or less successfully is the automatic teller machine (ATM). A basic transaction is easy to master. The machine prompts you to press a sequence of buttons that tell the ATM what you want to do. If you are a frequent ATM user, you have probably noted that some of the machines are more user-friendly than others. From my point of view, the friendlier machines are the ones with fewer buttons, and therefore fewer opportunities for error.

An ATM transaction involves both person-to-computer and computer-to-computer communication. The rules by which both parts of this communication are carried out, are called *protocols*. Protocols are crucial to computer-to-computer communications and have been researched actively. But if we are to make the most of our computer-oriented society, we must pay more attention to the person-to-computer protocols. We need to make the rules of discourse as much as possible like the person- to-person protocols that we learn from infancy. Only then will computers become truly accessible.

Computing and Communications

As computers have proliferated, the need arose to transfer data from one to another. It is easy to think of examples where interconnecting computers over short, medium, and long distances can be beneficial in normal business situations. In a common example, a manager may choose to interconnect the many computers in his office with a *local-area network (LAN)* so as to permit the rapid exchange of documents or files electronically. On a more mundane level, even the amateur computer user can buy a device called a modem that lets him connect his computer to a telephone line to provide access to stock market or other data located in a computer anywhere in the world.

The intimate relationship between computing and communicating is evident in the everyday ATM transaction. An ATM is itself a small computer that can interpret commands from a customer and then exchange information with a large, centrally located computer than stores the customer records. Most importantly, the ATM can dispense cash upon the authority of the bank's computer.

Several processes are implied by a conversation between a customer and an ATM. Some of the processes involve communication, and some involve computing. In Figure 1-3 we'll trace through part of a transaction involving a user (we'll call him Sam), the ATM, and the computer and dissect it into its major communication and computation processes. The transaction begins when Sam inserts his card into the machine, telling it who he is through his identification (ID) number stored on the card. The ATM responds with a request for Sam's password; having received this, it requests the type of transaction desired. Once Sam has entered the appropriate information, the ATM sends it on to the bank computer, where the password and Sam's bank

```
Sam: Inserts card
ATM: Reads user ID number on card
```

Communication—User to ATM

```
ATM: Displays request for password
Sam: Reads display
```

Communication—ATM to User

```
Sam: Enters password
ATM: Receives password
```

Communication—User to ATM

```
ATM: Requests nature of transaction
Sam: Requests withdrawal of $100

ATM: Sends ID, password, and withdrawal request
     to computer
Computer: Receives ATM transmission
```

Communication—ATM to Computer

```
Computer: Compares ID and password with its files
```

Computation

```
Computer: Transmits OK or not OK
ATM: Receives computer transmission
```

Communication—Computer to ATM

```
ATM: Interprets computer transmission
```

Computation

Figure 1-3 Communication and Computation Processes in an ATM Transaction balance are checked. If all is in order, the computer instructs the ATM to give Sam the money. If not, the withdrawal is denied.

It is clear from this example that to transport symbols we must manipulate them, and to manipulate symbols we must transport them over short or long distances. While we are all free to be purists and dissect every complex process into communication and computation, this is hardly worthwhile. There are certain things that relate to communication in a very narrow sense: the science and engineering of transporting information reliably and economically. There are also things that relate to computation in a very narrow sense: computational mathematics and the engineering of low-cost, fast, and reliable computing circuits. But once we leave these specific realms, the two merge into a broad, new field. Various names have been suggested for this symbiotic union; *compunications* and *telematics* are examples. Neither of these has caught on, and I am content to use the term *information technology*. Even though this field is still in its infancy, it has already influenced our daily lives, and will likely to continue to do so to a still greater extent.

Cultural Roots

Electrical telecommunications preceded automatic computing by close to a century. By the time the digital computer came along, analog telecommunications was a mature field. It has evolved slowly from analog to digital, but has always maintained its parental roots.

Computer engineers, seeing the need to network computers pressed for digital communication capabilities. In some cases, the communications professionals were not as responsive as they might have been, so many computer engineers became communications engineers in their own right. The first computer networks were developed by computer engineers, rather than by traditional communicators.

These same cultural divisions were present in academia, where departments of computer science were often established independent of the electrical engineering departments that housed the communicators. But after a time, the divisions began to break down as technical advances in both fields made it clear that this division was not only artificial, but counterproductive. Today, the trend is back toward combining the two in unified electrical engineering and computer science departments. This permits the maximum benefits to be derived from the synergy between the traditional disciplines.

Data Transfer

We can obtain greater insight into the computer/communications relationships by reexamining the basic structure of a computer. All of a computer's main elements store information, and the computer works by transferring information from one element to another. However, this process does not imply anything about the physical location of the elements. Every computer has some of those elements located in a single box. But it may also have some elements located elsewhere. Whenever this happens, we move from data transfer *within* a computer to telecommunication *between* remotely located computers.

What is the difference? While there is none at the conceptual level, at a practical level there are major differences in cost and accuracy.

As long as the entire computer is contained in a small space, we tend to think of the process of transferring information from one place to another as virtually trivial. In fact, we don't normally think of it as communication at all. But now suppose that the computer is in New York and a person in San Francisco transfers data to and from that computer using a keyboard. Communication is a consideration in this case, because an additional outlay of funds is required. But, there is no difference between the two cases in principle.

Let us now make the problem more complex and suppose that a large memory file containing millions of bits of data is to be transferred from a computer in San Francisco to the New York computer. We want the data transfer to be accurate, and we want it to take place within a short time interval. If the memory file were in the same room as the computer, there would be little concern. But long-distance, highly accurate, high-speed data transfer can be expensive.

Until the development of the computer, telecommunications technology was dedicated to the transfer of analog information over telephone wires and by radio broadcasting. As computers became more important, the need arose to exchange data with computers over distances. This led to the development of technology applied to finding ways to ensure that the communication was fast enough, accurate enough, and cheap enough for the application. It is interesting to note that information theory describing telecommunications from this point of view was formulated in the late 1940s, the time that the first digital computers were developed, well before anyone could have recognized the ultimately close relationship between the two fields.

Computer Resource Sharing: Benefits and Hazards

The first primitive digital computers could serve only a single user at a time. After awhile, computers became sufficiently powerful to be able to serve more than one user at a time. These *time-sharing* computers took advantage of the fact that the in/out units were much slower than the arithmetic-logic unit. While one user program was inserting or extracting data, the otherwise-idle arithmetic-logic unit could be servicing another user program. This was efficient, but it introduced another problem: if several programs reside in the computer simultaneously, then it is easy for one user to read data in another user's program or, even worse, to modify data in data files belonging to someone else. The possibilities for mischief are clear.

Computers have evolved from single-user systems to multiple-user systems and back again. The same technology that made time-sharing possible on large computers led to the minicomputer which initially served a single user. Further advances led to time-sharing on the minicomputers and to the development of still smaller single-user computers. Today, with computer networking, even single-user personal computers are effectively time-shared.

What can be done to eliminate or at least mitigate the problem of user interference? At the most elementary level, computer managers have introduced controls to make sure that only authorized users can gain access to systems. The ATM password system is a familiar example. At another level, the communications links between computers can be scrambled or encrypted to keep wiretappers from reading the confidential data being transferred between computers. These measures are vital when it comes to national security or to financial transactions by institutions such as the Federal Reserve.

Inside a computer there are certain steps that can be taken to protect the hardware from unauthorized interference. But the problem of protecting a program or its data files from an unauthorized user is exceedingly difficult. The Department of Defense has sponsored research into how to erect barriers between programs resident in the same computer at different security classification levels. These techniques would guarantee that a person at a lower classification level could not access data at a higher level. While it is possible to provide some protection through software barriers, guaranteeing the kind of absolute protection needed for national security is extremely difficult. Thus, for the most part, security musts be achieved by other, less convenient techniques.

Now that so many computers are interconnected, the phrase *computer virus* has entered the vocabulary. A computer virus is a program that can be surreptitiously inserted into all the computers interconnected by a computer network. (Even if your computer is not on a network, it can still be infected by a program distributed on diskettes.) Once inserted, the virus can infect the computers in any number of ways. In the extreme, it can erase all of the memory files resident on the computer. On a less serious level, it can simply display a taunting message about how easy it is to break into the computer. Whether the damage caused by a virus is fatal or merely debilitating, the whole subject of computer security has a serious dampening effect on the great benefits that shared computer assets can have for our society.

2

Telecommunications

We are now prepared to begin a more detailed discussion of telecommunications. The word *telecommunications* signifies the transfer of information over long distances. The concepts of analog and digital apply to telecommunications just as to computers. Modern analog telecommunication, as exemplified by telephone calls and radio and television broadcasts, uses signals that are, in some way, proportional to the audio or video information being sent. In digital communication, the information is in the form of numbers. When the information to be sent is computer data, the digital representation is straightforward. But when the source of the information is continuous, e.g., audio or video, the information must first be converted to a string of digits before it can be sent. Like digital computing, digital telecommunications has ancient origins, in smoke and drum signals. "One if by land, two if by sea," is the way that

Longfellow described the digital signals used to inform Paul Revere of the route of the British in 1775.

The *accuracy* of the communication is measured by the faithfulness of the received information to the transmitted information. In any practical telecommunications medium, there are disturbances of various kinds that distort the telecommunications signals. The greater these disturbances, the more difficult it is to achieve some desired degree of accuracy. With analog telecommunications as with analog computing, each increase in accuracy is successively more difficult to achieve. The essential advantage of digital telecommunications and one of the reasons for its increasing use is that any desired degree of accuracy is obtainable no matter how high the distortion.

Data and Information

What is information? Information has many shades of meaning. We tend to use it interchangeably with the word *data*; dictionaries are imprecise about the distinctions. Nevertheless, even in imprecise, every-day language, the word *data* implies a collection of words or numbers, the meaning of which has yet to be determined. The word *information* implies that the words and numbers have been processed in some way to extract some meaning or significance.

Digital communication is concerned with the transfer of digits from one place to another. When do these digits represent information, and when do they simply represent data? An example drawn from the computer structure outlined in Chapter 1 will illustrate the point. The control unit of a computer sends command signals throughout the computer. Imagine a primitive arithmetic-logic unit that can only add and subtract. It decides which to do based upon direction from the control unit. In this primitive computer, the control unit must be able to communicate just one of two things to the arithmetic-logic unit, "add" or "subtract." Based upon this statement, the arithmetic-logic unit takes the appropriate action. The control unit therefore sends information to the arithmetic-logic unit with this command. How much information? Since the transmission commands one of two things, a single binary digit will suffice to carry the information: it might send a 1 to designate *add* and a 0 to designate *subtract*. Thus we say that 1 bit of information has been sent. The word *bit* does double duty; it stands for a binary digit and also for a unit of information.

This might lead you to think that a binary digit always carries one bit of information. This is not necessarily so. Now imagine that the control unit can send only addition instructions and that the arithmetic-logic unit knows this ahead of time. In this case, no information is transferred when the control unit sends an addition command, because the control unit is telling the arithmetic unit something it already knows. This transmission consisted of a bit of data that carried no information and therefore needn't have been sent at all.

But that's an extreme case. How about a more realistic intermediate case in which the control unit sends addition instructions three-quarters of the time and subtraction instructions one-quarter of the time? In this case, the binary digit carries a fraction of a bit of information. Just how much is shown in Figure 2-1. The curve plots the amount of information carried by the binary digit against the likelihood that the digit is a 0 or a 1. At the left end of the curve, the transmitted digit must be a 0 and at the right end, it must be a 1. At these extremes, therefore, no information is transferred. In the center, it is equally likely that a 1 or 0 is being sent, and the maximum information of one bit is transferred. When the probabilities are unequal, a fraction of a bit is transferred; for the case at hand, about 0.8 bit.

Information Means Choice

Now imagine that there are 64 possible messages that can be transferred from one place to another. Both the sender and the receiver know the content of the 64 messages. The sender labels the messages

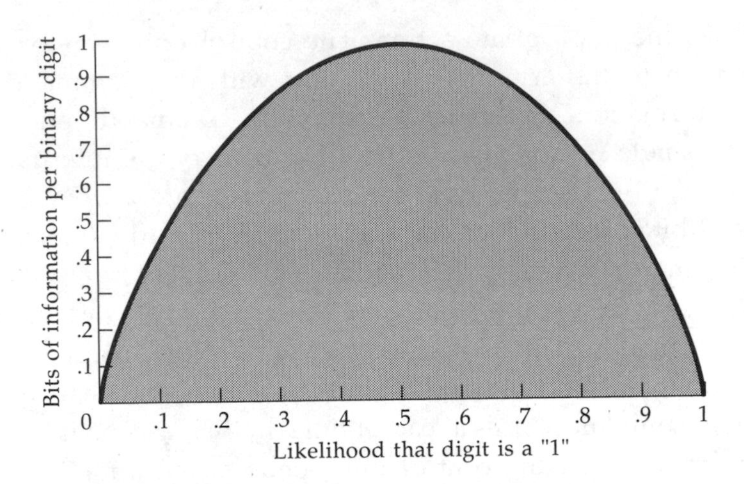

Figure 2-1 The Information Carried by a Binary Digit

from 0 to 63—in binary form, 000000 to 111111—and sends one of them using this binary label. It doesn't matter how simple or complex the messages are, provided that both the sending and receiving ends know their identity. One message might designate the entire Bible; a second, the complete works of Shakespeare; etc. The job of the recipient is to distinguish among the 64 possible labels to identify the correct one. How much information is transferred? As before, that depends upon how likely it is that the sender will transmit any particular message. If they are all equally likely, then the amount of information being transferred is the full 6 bits. At the other extreme, if one message—say 000000—is always sent (i.e., the likelihood of sending any of the others is zero), there is no need to send the message at all and no information is transferred. For any other distribution of message likelihoods, the amount of information transferred is between 0 and 6 bits, following a curve like that shown in Figure 2-1.

Now if the *information* transmitted is much less than 6 bits, then transferring 6 bits of data is wasteful. What can be done about it? Samuel F. B. Morse addressed this problem in a very intuitive way.

Morse code uses sequences of dots and dashes to represent the letters of the alphabet. Morse knew something about the relative frequency with which the different letters are used in ordinary English, and he designated the most frequently used letters with the shortest sequences and vice versa. Thus, the letter *e* is designated by a single *dot* and the letter *q* by the relatively long sequence *dash-dash-dot-dash*. Following his example, if we know the relative frequency at which the messages are sent, then we can take our 64 message labels, encode the most likely with short codes and the less likely with longer codes, and reduce the average number of transmitted bits to something close to the actual amount of information being transferred.

We have, in effect, defined information transfer in terms of the amount of *choice* being transferred. In a choice between one of two equally likely messages, a single bit of information is transferred. In a choice between one of ten equally likely messages (one decimal digit), between three and four bits of information are transferred.

Note that this definition of information has nothing to do with the inherent information content of the data referred to by the labels. Thus, only one bit is needed to distinguish between the transmission of Hamlet and Macbeth assuming that the receiver has perfect copies of each and needs only the information relative to the choice between the two. However, if one computer wants to send a copy of Hamlet to the other, the situation is completely different. In this case, all the text of the play would have to be converted to a sequence of binary digits. If this conversion were performed without regard to the letter or word frequency in the text, then 0s and 1s would occur with equal frequency, and the communicated information *rate* as defined above would be far in excess of the actual information *content* of the text. Taking into account some of the properties of the alphabet and the language would enable us to reduce the data rate toward something closer to the true information content. Taking into account the semantic content of the language would reduce the information rate still more. If we knew how to gain insight into this inherent information content, we could, in principle, encode the data so that many fewer digits would be required for the transmission, something closer yet to the true information content of the text.

This information-oriented approach points out a difference between short- and long-distance communication. If the distances are very short—say, within the computer itself—then the chance of a signal being identified incorrectly at the receiving end is extremely low. But if the distance is long enough to require, say, a telephone circuit, there is a chance that the message being transmitted will be corrupted enough to make the receiver misidentify the message. This corruption of the message introduces ambiguity in the receiver's decision process which is equivalent to reducing the information transferred by the data. In other words, if 6 bits of information were transmitted by the sender, something less than 6 bits would be received by the recipient. Suppose that 1 bit is lost in its entirety so that 5 rather than 6 bits are received. Then every received message can be one of two things, depending upon the identity of the missing bit. In other words, since 5 bits represent a choice of one of 32 things instead of one of 64, the result is a confusion factor of one in two messages. Since confusion of this sort is intolerable if it occurs too often, the communicator must either reduce the information transmission rate of the source or improve the communications medium to ensure that less information is lost.

The Transfer of Information

The primary goal of telecommunications is to preserve the information content of the transmission to the extent necessary for the application at hand; or, in other words, to transmit the data representing the information with an *accuracy* commensurate with the application. The essential virtue of digital communication is that its accuracy can be controlled with ease. The same goal exists for analog communication, although its accuracy is harder to quantify and control.

A second goal is *efficiency*. While its dictionary definition is "the production of a desired effect without waste," it has various definitions depending upon context. Thus, the more efficient your automobile engine, the more miles your car gets from a gallon of gas. But efficient doesn't necessarily imply cheap. If a highly efficient automobile engine costs a great deal more than a less efficient one, then the net cost to the customer may be greater despite the fuel savings. The solution to a problem may be both cheap *and* efficient, but not necessarily.

A scientist seeks to understand nature for its own sake, whereas a technologist or an engineer is one who applies scientific knowledge to practical ends. To state that the goal of telecommunications is to achieve accuracy and efficiency in the communication of information is to give you a scientist's perspective. The telecommunications engineer's goal is to use this scientific knowledge to develop or produce telecommunications components and systems that do an acceptable job at an affordable cost. Telecommunications scientists have developed a whole body of analytical techniques that improve the accuracy and efficiency of information transmission in the face of the distortions and noise that corrupt the telecommunications signals. Which of these techniques turn out to be of practical use is determined by an engineer in some specific context.

An example from the telecommunications world is illustrated in Figure 2-2. Suppose that a company must send large data files from a computer in San Francisco to one in New York using telephone lines. From the size of the data file and the time sensitivity of the data, the data-processing director has determined that the data must be sent at the rate of 9600 bits per second (abbreviated bps). Furthermore, this transmission must be very accurate—so accurate that, on the average, no more than one in ten million bits transmitted can be received incorrectly. The telecommunications manager can do the job in one of the two ways shown in the picture. In either case, a *modem* is used to convert the digits into signals appropriate for the telephone line.

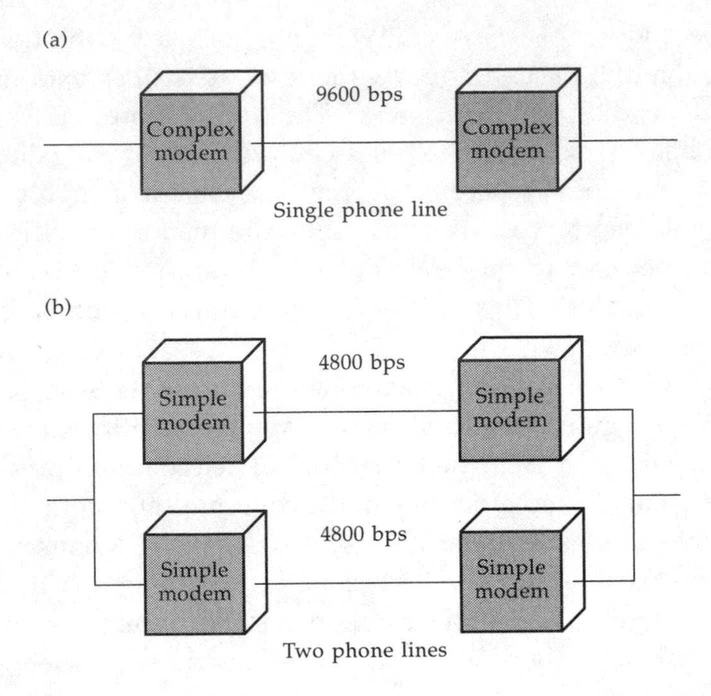

Figure 2-2 Communicating over Telephone Circuits

Modem, an abbreviation for *mo*dulator-*dem*odulator, has come into the vernacular in recent years in the context of personal computers. I have even heard it used as a transitive verb, as to *modem* data from one computer to another. As barbarous as this may sound to those who value the purity of the English language, it certainly makes its point. The meaning of the terms, modulator and demodulator, will become clearer later on.

In Figure 2-2a, the modems are sophisticated enough to support a rate of 9600 bps using a single phone line, while maintaining the desired transmission accuracy. In Figure 2-2b, the modems are simpler and cheaper, but can only meet the accuracy requirement at half the rate. Therefore, to meet the 9600 bps rate requirement, two phone lines are needed. The first case uses the phone line twice as efficiently as the second case. But since the telecommunications manager is interested only in cost, not in efficiency, he or she will choose the efficient solution only if the cost of the efficient modems is less than the cost of the less-efficient modems plus the cost of the additional phone line. Had this choice been presented to the communications manager as recently as the early 1970s, the economical choice would have been the less efficient configuration. But with modern integrated-circuit technology, the more efficient modem costs scarcely more than the less efficient.

Information Theory

The fundamental science underlying telecommunications, addressing and explaining in an elegant way the fundamental issues of efficiency and accuracy, is called *information theory*. Its foundations were laid in 1948 with the publication of two papers by Claude Shannon, then at the Bell Laboratories and later a member of the MIT faculty. In the years since this great work, many others have developed the mathematical techniques that have permitted practical application of his results, capitalizing upon the advances in component technology as in the above example.

Shannon was the first to formulate the notion of information as choice. He called it *entropy* because of its remarkable resemblance to the thermodynamic quantity by that name that indicates the degree of randomness of the motion of gas in a container or electrons in a metal. Others had recognized the relationship as well, but only Shannon was able to put everything together in such a coherent fashion. Jeremy Campbell, an English science writer, notes in his highly-acclaimed book, *Grammatical Man*, that a year before Shannon's publications, Norbert Wiener, a mathematician well known not only for his intellectual accomplishments but also for his eccentricities, was observed by Robert Fano, a younger colleague, to mutter "information is entropy" between puffs on his cigar while walking the halls of MIT.

While Shannon's ideas were developed solely for communications engineering, they have been very influential in other fields. To quote

Campbell, "Essentially [Shannon's] papers consisted of a set of theorems dealing with the problem of sending messages from one place to another quickly, economically, and efficiently. But the wider and more exciting implications of Shannon's work lay in the fact that he had been able to make the concept of information so logical and precise that it could be placed in a formal framework of ideas." And indeed, *Grammatical Man* addresses itself to the insights provided by information theory and the thermodynamic concept of entropy to such fields as linguistics, genetics, and cognitive processing.

Not everyone agrees with such a broad interpretation. Even Shannon, a very self-effacing, introverted man who published seemingly only under duress throughout his career, suggested in an article in 1956 that "information theory has perhaps ballooned to an importance beyond its actual accomplishments." Speculations on the ultimate importance of information theory are beyond my purpose. But it is Shannon's seminal contribution to communications in the narrow sense that are my concern, both because of its intellectual stimulation and for the foundation it provides for the practical business of telecommunications.

Energy and Noise

Transmission efficiency is defined as the transmitted information rate as compared to the maximum sustainable in a communications channel. It is now appropriate to ask what properties of a communications channel determine this maximum data rate. Let's look at an analogous problem: what factors determine how much water can be transported through a pipe from a pumping station? Two things seem evident: the greater the water pressure developed at the pumping station and the larger the diameter of the pipe, the greater the rate of flow should be. There also is the effect of the condition of the inner surface of the pipe.

If this surface is rough and uneven, the water will flow turbulently rather than smoothly, and this will inhibit the rate of flow of the water.

In communications, the amount of energy in the signals is analogous to the water pressure, and the noise level is analogous to the roughness of the pipe. Intuitively, the greater the signal energy and the less the noise, the greater the information rate that can be supported. The analog of the pipe size is the *bandwidth* of the communications channel. We will concentrate on signal energy and noise energy in the rest of this chapter, and will return to bandwidth later.

The effects of signal strength and noise are universal. The channel may be acoustic, electrical, or optical; it may be analog or digital. The important parameter in determining the communication characteristics is the ratio of the signal strength to the noise, rather than the absolute value of either one. The following example illustrates why this is so.

Jack and Jill meet every day for lunch in a picnic area in the center of a large park. They talk quietly as they eat. On this particular day, the park is empty. The only sound intruding is the barely audible traffic noise coming from the heavily traveled public roads bordering the park. The time has come to part. Jack reluctantly starts walking away; Jill remains. After a few moments, Jack remembers something that he had meant to say and turns around to tell Jill. He is now a few yards away, and he has to raise his voice a bit so that Jill can hear. She is now aware of the road noise because it competes ever so slightly with the reduced intensity of Jack's voice, but she can still understand him perfectly well. Once again, Jack turns around and continues walking. After another 20 yards or so, he has another thought, and turns around to catch Jill's attention. Now he must shout in order to make his voice heard over the traffic noise. In fact, Jill misses a few words and has to shout back to ask him to repeat his sentence. He does so, and Jill shouts her agreement. Jack's lunch break is almost over. He rushes along. Again he remembers something that he had intended to tell Jill. Again he turns around. But now he is about 100 yards away, and he has to shout as loudly as he can just to attract Jill's attention. Even so, she can barely hear him. The automobile noise that seemed so faint when the two were together is now loud enough to mask the reduced intensity of Jack's voice.

The example demonstrates the qualitative effect of signal energy and noise on the transmission and reception of sound. When Jack and Jill were at a normal conversational distance, Jack's voice level was high enough to be understood perfectly. The noise level relative to the intensity of his voice was negligible. When they were a few yards apart, Jack was able to raise his voice enough that Jill was able to hear and understand him well enough, although the traffic noise was now discernible. When separated by a greater distance, Jill had all she could do to make out Jack's words. She even missed a few. But when they were separated by the length of a football field, Jack couldn't shout loud enough to overcome even the low level of the traffic noise.

We can make an important observation from this example. As Jack moved away from Jill, the energy in his voice signals reaching Jill was reduced—in fact by the square of the distance separating them. This means that the acoustic energy decreased by a factor of 100 when he moved from a distance of 2 feet to 20 feet, and by another factor of 100 when he was 200 feet away. In contrast, the noise level remained the same. It just appeared to increase because of the reduced voice energy. This illustrates the earlier point that what is important is not the signal or noise energies themselves, but their ratio.

Another observation from this example is that acoustic communications have a very limited range. If Jack had had a megaphone he might have increased his conversational range a little, but ultimately distance would have defeated him. He would have been forced to walk back to within earshot of Jill or to wait until both were at home so he could call her on the telephone to complete the conversation.

To send information over long distances, it is essential to be able to boost the energy level of the communication signals enough to overcome the background noise. Practically speaking, this means using a medium that transports the signals electrically or optically rather than acoustically. The reason for this is that it is easy to *amplify* electrical signals and difficult to amplify acoustic ones. An *amplifier* is a device that boosts the energy in a signal. The most familiar example is the audio amplifier, which adds enough energy to the faint signal coming from a phonograph pickup to drive a loudspeaker. Where does that energy come from? From an *electrical power supply* and, ultimately, from your power company.

It is significant that there is no purely mechanical equivalent to the process of amplification. For example, the first phonographs were all acoustic. Singers had to shout into the equivalent of an inverted megaphone to produce enough energy to drive the stylus used to cut the record. The music was played back again with a megaphone-like horn. A great leap forward in music reproduction occurred in the 1920s with the use of the first electrical systems. The music to be recorded was first converted from acoustical to electrical form using a microphone, was amplified electronically, and then was reconverted to mechanical form to cut the record. For playback, the weak acoustic signals were converted to electrical form, amplified electronically, and then reconverted to acoustic form in a loudspeaker. The multiple conversions from acoustical to electrical form and back again do take their toll in musical fidelity. But this is a small price to pay for the revolution in sound quality brought about by the use of electrical techniques.

Defining Energy

We are all consumers of energy in different forms. When we burn a 100-watt light bulb for one hour to illuminate a room, we consume 100 watt-hours of electrical energy. Should we burn the bulb for 10 hours, we would consume 1000 watt-hours or 1 *kilowatt-hour*. This is the unit by which we buy energy from our utility companies. The definition of energy is related to that of *power*, defined as the rate at which energy is generated or consumed. The unit of power is the watt. Electric light bulbs and toasters are described by their power consumption. Their energy consumption is then given by their power consumption multiplied by the length of time that they are used.

While the kilowatt-hour is a very convenient unit of energy, it is not the fundamental one. That distinction belongs to the *joule*, the name given to a watt-second, the energy consumed by a one-watt bulb in one second. A kilowatt-hour is therefore equivalent to 3,600,000 joules (1000 watts in a kilowatt, and 3600 seconds in an hour). Even though the joule is the basic unit, utility companies do not use it because it is too small to be convenient. For if energy were sold by the joule, its price would be in the vicinity of .000003 cents, the equivalent of 10 cents per kilowatt-hour. Should the cost of energy ever escalate wildly, you might find the utility companies pricing their energy by a smaller unit, like the kilojoule, to soften the apparent blow on the consumer, following the lead of the European gasoline companies who typically price their relatively expensive gasoline by the liter rather than the larger gallon or equivalent metric unit.

The definitions of the watt and the joule are not electrical, but are derived from the branch of physics called mechanics. The watt was named after James Watt, the inventor of the first successful steam engine, whose work was fundamental to the practical generation of mechanical energy and had nothing at all to do with power in electrical form. The energy of a body in motion, called its *kinetic energy*, is defined as one-half the mass of the body multiplied by the square of its speed. Thus, a subcompact car weighing 2200 pounds or 1000 kilograms moving at a speed of 60 mph has a kinetic energy of about 360,000 joules, or about 1/10 of a kilowatt-hour. The automobile's engine burns gasoline to provide kinetic energy to the car, which compensates for the frictional losses that continually convert some of the kinetic energy to heat. When the driver applies the brakes to stop the car, this kinetic energy is converted to the thermal energy of friction between the tires and the road. If the car were to crash into a stone wall, all this kinetic energy would again be converted into heat, enough to burn a 100-watt light bulb for an hour.

This discussion illustrates one of the fundamental laws of physics, the *law of conservation of energy*, which states that energy may be neither created nor destroyed, only converted from one form to another. A light bulb converts electrical energy into optical and thermal energy. The kinetic energy of the automobile is converted into thermal energy after the crash.

Communication Requires Energy

When you talk into the telephone, the electrical signals from the microphone contained in the handset are amplified to compensate for the energy they lose in traveling to their destination over hundreds or thousands of miles. The transmission media within the telephone networks are electrical or optical, using either cables (telephone lines) or propagation through free space (radio signals). The signals are generated at the source at a power level high enough to reach the destination after suffering degradation in the medium. Cables, whether optical or electrical, introduce dissipative losses that limit the range that a signal can travel. To compensate for these losses, amplifiers are installed at appropriate locations along the route to boost the signal strength. The degradation in radio media is caused by the spreading out of the signal with distance and, in some cases, by dissipative losses as in the cables. Radio and wire-line transmitters amplify signals so that they can go virtually any distance and still be able to overcome the ambient noise. As a result, we are able to talk very comfortably over very long distances.

How Much Energy?

Just as it takes a certain amount of energy to generate light, to heat water, or to propel an automobile, so it takes a certain amount of energy to communicate information. Since the ratio of signal energy to noise energy is the factor that determines the quality of the sound reception, the number of joules of signal energy depends upon the number of joules of noise energy present. In the earlier example, Jack's voice signals had to be strong enough to counteract the distant auto noise. In a hypothetical situation in which there is no noise at all, there is a threshold value of signal energy required for the receiver to perceive that a signal is present. If you were in a perfectly noiseless room, you could hear even the faintest whisper, provided its energy exceeded the minimum level necessary to trigger the fundamental physiological processes of hearing. But we can safely ignore this hypothetical case. In all practical cases, noise of one kind or another is always present, and the signal energy usually must be at least some 10 times as large as the noise energy for satisfactory reception. Shannon showed that there is an ultimate limit, not achievable in practice, in which the signal energy can be just under 70% of the noise energy.

Noise

Noise is a foreign signal of any origin that competes with the communications signal that you want to hear. The optical signaling scheme used to inform Paul Revere of the route of the British in 1775 might not have succeeded had there been enough stray light from other sources to mask the lantern signals from Boston's Old North Church. The traffic noise that disturbed Jack and Jill is an example of acoustic noise. Another example is the boisterous conversation from a nearby restaurant table that interferes with your business lunch. Noise can also be electrical, and it can come from many sources. You may hear it when using the telephone. When listening to the radio, you sometimes hear interference from *static*, the name commonly given to atmospheric noise caused by electrical storms. You may see "snow" on your TV set, particularly if you live in a fringe area far enough from the broadcasting station that the signal level is diminished. If you replace your rabbit-ears antenna with a more powerful roof-top antenna, you are increasing the amount of broadcast signal that enters your TV. The snow will diminish because the signal level has been increased relative to the noise level. Signals can also interfere with themselves. The "ghosts" on your television screen are the result of multiple reflections from objects in the path of the TV signal. Each of these reflections is delayed by a small amount relative to one another, resulting in multiple copies of the same signal entering the antenna at slightly different times.

Some noise sources are easy to identify: e.g., a lightning bolt or a loud voice. But some are more difficult because they are the combined effect of a large number of noise sources. The traffic noise intruding on Jack and Jill is the sum of a large number of noise sources. Similarly, the needle scratch from an old record is the result of a very large number of fluctuations caused by abrasions of the record, and the soft background hiss that you hear on the telephone or radio is the result of a very large number of interfering signals. Whenever the number of interfering signals is large, the most useful way to describe their effects is to address their aggregate behavior. Noise of this kind is called *thermal noise*.

Thermal Noise

Electrons in all materials are continually in motion, often colliding with one another and with the atomic nuclei. In each of these collisions, energy is transferred from one particle to another. Any pool player has observed that when one ball hits another head on, the first ball stops and the second moves with almost the speed of the first ball before the collision. In other words, aside from a small amount of frictional loss, all the kinetic energy of the first ball is transferred to the second in the collision.

If we were to plot the velocity of an individual electron as a function of time, we would obtain a zigzag picture similar to that shown in Figure 2-3. By velocity, we mean not only how fast the

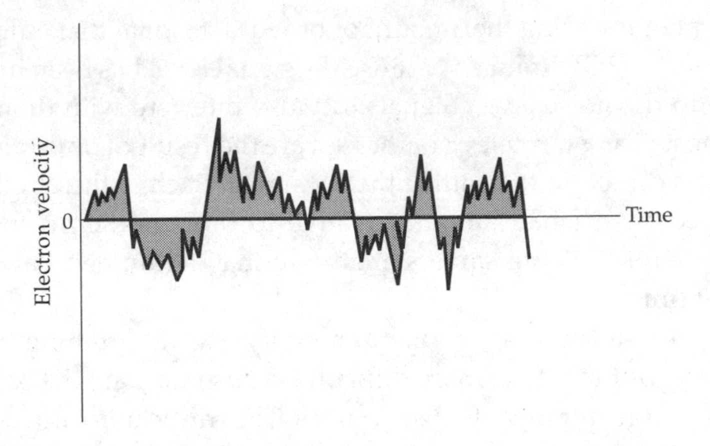

Figure 2-3 Random Electronic Motion

electron is moving (its speed) but also in which direction. For example, we might define positive velocity to be that of the electron moving from left to right and negative velocity to be that of the electron moving from right to left. As you might expect in such random motion, each electron will be moving backward as much as it is moving forward, and its average velocity should be zero.

Another important thing to know about the motion is how much of the time the electron is moving slowly and how much rapidly. To arrive at this information, we imagine another hypothetical experiment. Instead of concentrating on the wanderings of a single electron, we make simultaneous measurements of the velocities of a very large number of electrons, all of which behave the same way statistically, and plot the result as in Figure 2-4. The curve has the bell shape that is called *normal* or *Gaussian*. While most of the electrons have low speeds, some smaller number in the "tails" of the curve have higher speeds. Note also that the curve is symmetrical about zero, agreeing with our previous observation that the average velocity is zero.

In a communications receiver, the electrons in all the constituent components are in random motion, and this motion is superimposed

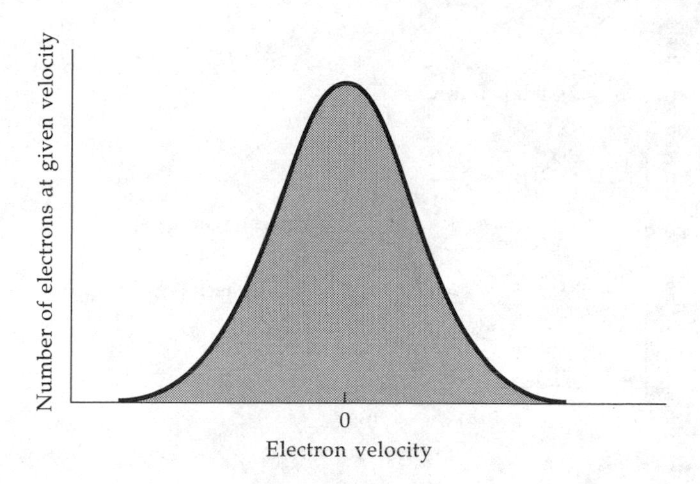

Figure 2-4 Velocity Distribution of Electrons in a Material

upon the deterministic (nonrandom) motion caused by signals. This random motion of the electrons is the source of thermal noise (sometimes called Gaussian noise, from the shape of the curve in Figure 2-4).

We call this noise thermal because the average electron kinetic energy is determined by the temperature. The higher the temperature, the higher the energy. The temperature at which all motion stops defines the lowest possible temperature, called *absolute zero*. This is the basis for the *absolute* or *kelvin* temperature scale. It is the same as the familiar celsius scale, but displaced by 273.2 degrees, so that zero degrees celsius is 273.2 kelvin or K.

The curve of Figure 2-4 was measured at a particular temperature. Figure 2-5 shows several similar curves at different temperatures. The lower the temperature, the thinner the curve. In the limiting case at absolute zero, the curve degenerates to a line at zero velocity, reflecting the fact that all the electrons are motionless. The average value of the velocity in each case is still zero, since an electron is just as likely to be moving forward as backward, regardless of the temperature. A more interesting quantity is the average value of the *square* of the velocity, which is proportional to the average kinetic energy of the electrons.

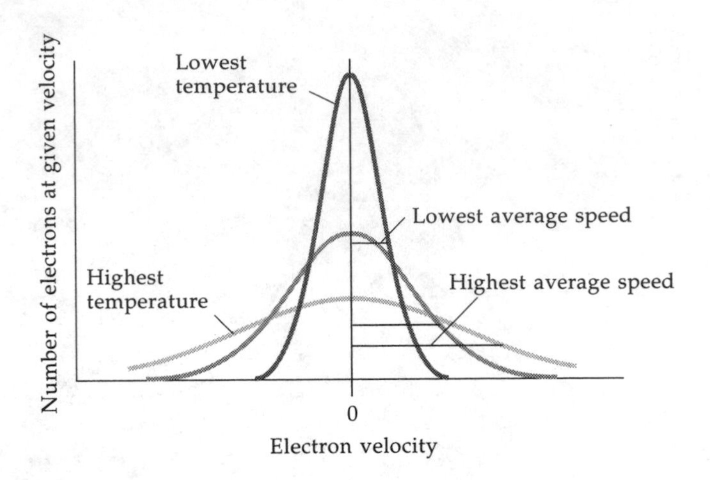

Figure 2-5 Electron Velocities at Different Temperatures

The width of each of the curves is proportional to this average energy. Since the random electronic motion is the source of random noise, the noise energy is this average electronic kinetic energy. It is proportional to the temperature and is zero at absolute zero.

Our understanding of thermal noise is derived from the branch of physics called statistical mechanics, which describes the properties of large ensembles of particles (molecules, atoms, electrons, etc.) as opposed to the properties of the individual particles. From this point of view, the properties of a large ensemble of electrons in a material are similar to the properties of a large ensemble of oxygen molecules in a container. Early in the nineteenth century, it was discovered that the average energy of a molecule in thermal equilibrium is proportional to the absolute temperature, where the proportionality constant is 1.38×10^{-23} joule per Kelvin. (The notation means 1.38 with enough zeros in front to move the decimal point 23 places to the left.) Like the speed of light and the electrical charge on the electron, this is a fundamental constant of nature, known as Boltzmann's constant after the famous nineteenth century German physicist Ludwig Boltzmann. Thus, at normal room temperature of approximately 27° C (300 K) the thermal noise energy is 4.14×10^{-21} joules. This is a very small number, and its absolute value won't mean very much to you at this point. But it is a very important number because it represents the fundamental limitation on the amount of energy required to communicate.

Digital Communications

The dependency of communications performance on the ratio of signal energy to noise energy is universal, applying equally to analog and digital communications. The earlier example was of analog communications. We will now focus on digital data transmission.

Figure 2-6 shows an example of a binary communications system that might use any of the transmission media that have been mentioned. To send a 1 in this electronic system, we transmit a rectangular pulse of energy in some time interval, and to send a 0, we transmit nothing during the time interval. At the receiver, the transmitted pulse is corrupted by thermal noise. A 1 is received as the pulse of energy with noise superimposed upon it. A 0 is received as a pulse of noise only. The receiver adds up all the energy received in the appropriate time interval and attempts to differentiate between 1s and 0s.

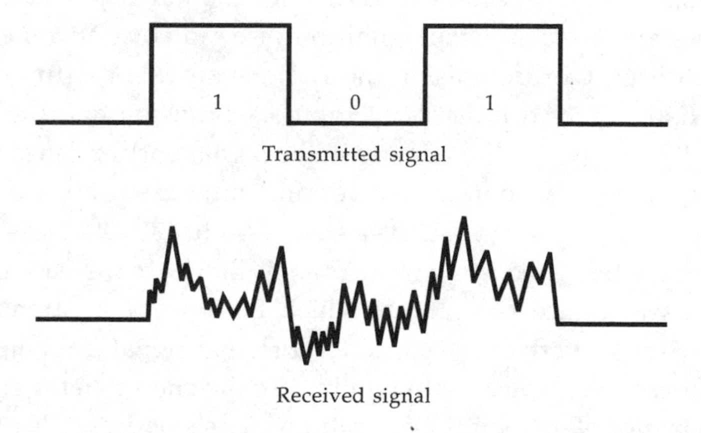

Figure 2-6 Digital Signals in the Presence of Noise

What does the term *pulse of energy* really mean? Whenever you switch an electric light bulb on and off, you are generating a pulse of energy. If a 100-watt bulb is left on for 1 second, the energy drawn from the power company is 100 watt-seconds, or 100 joules. The pulse of energy representing the 1 in Figure 2-6 is analogous. The transmitter generates some number of watts for some time interval. Thus, if the transmitter power level is 1 watt and if the pulses are 1 second long (corresponding to a data transmission rate of 1 bit per second), then each pulse contains 1 joule of energy.

The noise fluctuations can either add to or subtract from the signal. If the signal is much larger than the noise, it doesn't matter very much how the noise behaves; the receiver will be able to distinguish between 1s and 0s almost all the time. However, if the signal and noise energies are close to the same strength, then fluctuations in the noise can cause a 0 to look like a 1 and vice-versa. When this happens, the receiver can misidentify the signal. Note that even when the signal-to-noise ratio is very high, errors can still occur, albeit with a very low probability. This is because the noise is a statistical phenomenon and there is always the chance that an improbable noise peak can occur that will cause the receiver to make an error.

Suppose that with a transmitted power level of 1 watt and a data transmission rate of 1 bps, the data transfer is not reliable enough. It is then necessary to increase the signal-to-noise ratio by either increasing the signal energy or decreasing the noise energy. Let's suppose that there is nothing we can do about the noise; this is often the case in practice. Therefore, we must increase the signal energy. Since the signal energy depends upon the power and time duration, we can increase either one or the other, or perhaps both. Either approach will have the drawback of costing more money. But if we increase the time duration, we reduce the rate at which bits are being transmitted. Figure 2-7 shows both approaches. The original signal from Figure 2-6 is reproduced in Figure 2-7a. In Figure 2-7b the energy in the signals is doubled by doubling their time durations to 2 seconds, while retaining the original power level of one watt. In Figure 2-7c, the original power

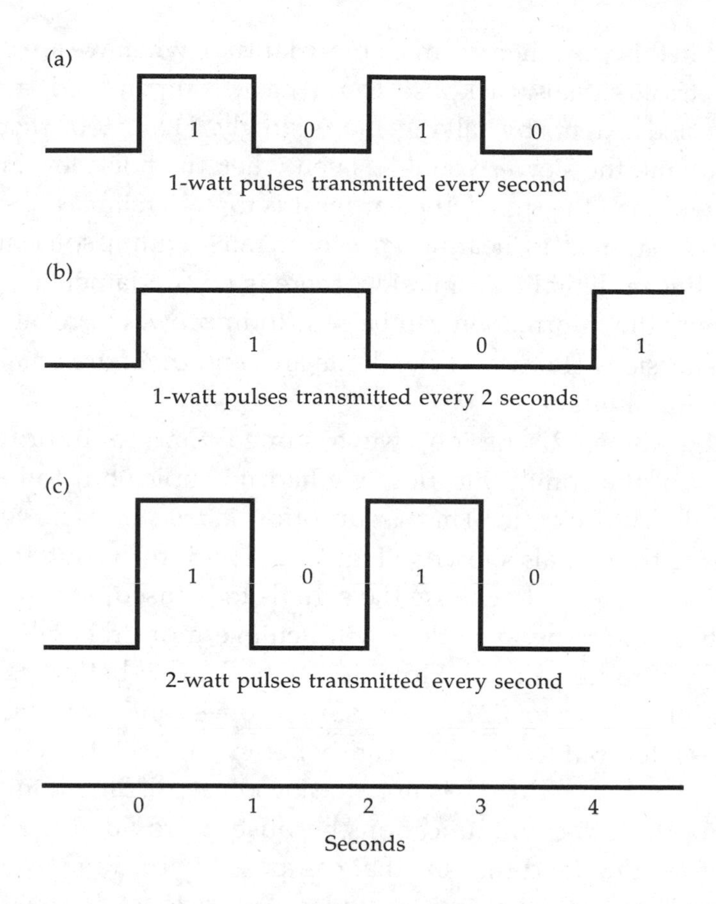

Figure 2-7 Doubling the Transmitted Energy

level is doubled and the original pulse width is retained. In either case, doubling the signal energy for the same noise energy will decrease the chance that the receiver will make errors in identifying the received pulses.

Going back to our example of Jack and Jill in the park, as Jack walked away from Jill, the level of sound reaching Jill decreased. Jack was able to compensate for this to some extent by shouting, the equivalent of increasing the signal power as in Figure 2-7c. While human physiology limits the amount of power increase that can be achieved in this way, there is much more latitude when we are dealing with electrical signals. Jack also could have compensated, again to some limited extent, by talking more slowly. Here, too, there is a practical limit: the slower you talk, even when the noise level is very much lower than the signal, the harder it is to understand the speech. We are accustomed to hearing speech at rates within some normal bounds. But in digital transmission, there is no fundamental limit to how slowly the information can be sent to improve the reliability of the transmission. The slower the digits are sent, the more reliable the transmission should be.

In the above discussion, we assumed that to improve the reliability of the communication, we had no choice but to increase the signal-to-noise ratio. That assumption is true only if we insist upon using the signals shown in Figure 2-7. As it turns out, there are many different ways to choose the signals to be used, and some are better than others because they will achieve more reliability for a given signal-to-noise ratio. How do we find these better signaling schemes? It's easy to find some simple ones, and we will show examples later. But to do significantly better—in fact, to do the best that can possibly be done—is not an easy job at all. Shannon's work is so important to communications because it provided the insight needed to understand the best that can possibly be done. And even if the best is not achievable, it provides the yardstick for evaluating just what is achievable.

Shannon called the maximum possible rate at which error-free performance can be achieved through a communications channel, albeit using infinitely long signals, its *capacity*. It is a mathematical concept used to define the best possible performance, even if this best possible performance is not realizable in any practical sense. To take just one example, the capacity of a typical telephone line is somewhere around 40,000 bps. The example shown earlier in Figure 2-2 used rates of 4800 and 9600 bps, but even these rates, while often achievable, are still higher than the 2400 bit-per-second rate that is most often used. It is very difficult to achieve rates above 20,000 bps with very high accuracy. And since we know that going higher than 40,000 bps is impossible, we know better than to try.

3

Sound, Music, and Speech

All tastes and levels of sophistication respond to music. We do this because music moves us in inexplicable ways, as it has from the earliest times. There is no more eloquent tribute to the importance of music in the ancient world than the closing verses of the Book of Psalms:

> *Praise Him with the blast of the horn;*
> *Praise Him with the psaltery and harp;*
> *Praise Him with the timbrel and dance;*
> *Praise Him with stringed instruments and the pipe.*

But it is virtually impossible to put into words why it is that music affects us so. As Leonard Bernstein has written in the introduction to his widely-acclaimed book *The Joy of Music,*

> Ultimately one must simply accept the loving fact that people enjoy listening to organized sound; that this enjoyment can take the form of all kinds of responses from animal excitement to spiritual exaltation... .

This chapter examines the physical characteristics of the sounds that make up music and speech, rather than the spiritual attributes of the music. Fortunately explaining the issues underlying the physical nature of "organized sound" is far easier than explaining its esthetics. Thus my task is far easier than Leonard Bernstein's.

The field of psychoacoustics, which relates the way we perceive sounds to their physical characteristics, attempts to bridge part of this gap between the physical and the aesthetic. Some composers have taken advantage of this work by using computers to generate sound patterns that create unusual psychoacoustic effects. These composers are freed from the restrictions accepted by those composers who work within the limitations of the traditional instruments. But I am afraid that the psychoacousticians will readily agree that they are still a long way from understanding the sounds that a musician calls "organized." As John Pierce concludes in *The Science of Musical Sound*,

> . . . to succeed, new music must really be heard in the sense that the composer intended, must be understood, must hold the interest of and move the listener. Here an understanding and exploration of the science of musical sounds can help. The rest only talent or genius can supply.

Sound Production

All sound is caused by vibrations. Sound, musical or not, originates when something physical vibrates: a steel wire, a membrane, an enclosed column of air. This vibrating source, in turn, induces vibrations in the molecules of the surrounding air. The air vibrations propagate out from the source and induce vibrations in the ear of the

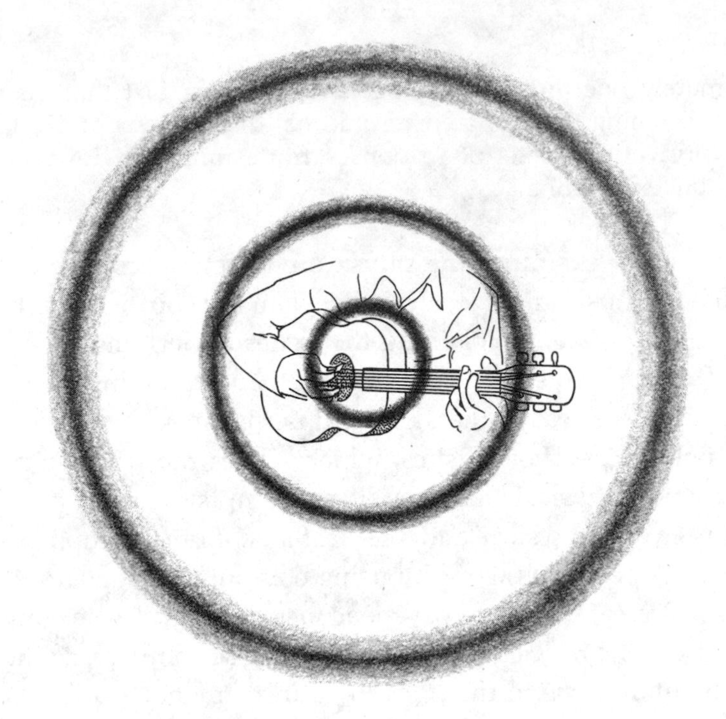

Figure 3-1 Sound Propagation (From J. R. Pierce, *The Science of Musical Sound*, Scientific American Books, New York, 1983)

listener. Stimuli are then sent to the brain to produce the sensation that we call hearing. I doubt that Shelley was thinking of mechanical vibrations when he wrote:

Music, when soft voices die,
Vibrates in the memory

but that does not render his verses less appropriate.

Sound travels in the form of waves. The characteristics of sound waves are similar to those of water waves. When you drop a pebble into a still pond, you can see the motion of the characteristic peaks and troughs of the waves as they ripple outward over the surface of the pond from the impact point. Of course, sound waves are as invisible as the air molecules of which they are constituted. If we could see them, they would resemble the alternating regions of higher and lower air molecule density emanating from the guitar music source in Figure 3-1, instead of the peaks and troughs of water waves. The louder the sound, the greater the density differential. They also propagate in three dimensions rather than two as do water waves. Clapping one's hands in the middle of a room, creating an acoustic impulse analogous to dropping a pebble in water, sends the density variations of the sound wave out radially from the impulse at a speed of about 1100 feet per second.

When the source of the initial disturbance is complex, the resulting waves are similarly complex. Waves that are generated by children splashing around in the water are very hard to characterize. Similarly, the vibrations made by an orchestra playing complex music generate sound waves that are hard to characterize as simple patterns of high- and low-density air molecule vibrations. But there is a structure to musical sounds, because these sounds, however complex, are combinations of simple sounds. This fundamental property gives us the language with which to discuss not only musical sounds but also the characteristics of digital signals.

Musical Instruments

Every musical instrument has some initial source of vibrations. Perhaps the most versatile of these primary vibrating sources is the stretched string, which may be plucked, bowed, or struck depending upon the particular instrument. The primary vibrations then induce vibrations in the rest of the instrument. A violinist may start the strings vibrating by plucking them or drawing the bow across them. The characteristic violin sound that we recognize results from vibrations in the body of the instrument induced by the string vibrations.

Similarly, a piano string starts vibrating when struck by a felt-covered hammer actuated by striking a key. These string vibrations are picked up by the sounding board of the piano and by some of the other strings and create the characteristic sound of the piano. The sound in wind and brass instruments originates in a vibrating column of air. In a clarinet or an oboe, a vibrating reed stimulates vibrations in a column of air inside the instrument. The same is true of a brass instrument—e.g., a trumpet or trombone—where the vibration of the lips into a mouthpiece provides the initial stimulation.

We can learn more about the characteristics of musical vibrations by examining the vibrating string a little more closely. Take the guitar, for example. Each string is stretched tightly between two points. Plucking the string causes it to vibrate back and forth. How rapidly it vibrates depends primarily upon three characteristics of the string: how long it is, how heavy it is, and how tightly it is stretched. Although the string vibrates too rapidly to observe its rate optically, you can "observe" it acoustically by listening to its pitch—since the greater the vibration rate, the higher the pitch. You can easily observe that the lower pitched strings are thicker than the higher pitched strings. Note, too, that one end of the string is held fixed while the other end is wound around a peg. The musician tunes the instrument by tightening or loosening the string with the peg; the greater the tension in the string, the higher the pitch. Finally, you can observe that the musician plays the various notes of the scale by pressing firmly on the string at different positions along the neck of the guitar with the left hand while plucking the string with the right hand. The action of the left hand causes the string to vibrate between the fixed pin and the position of the left hand. Thus the pitch is adjusted by varying the length of the string—the shorter the string, the higher the pitch.

A string can vibrate at many different rates, each derived from a particular vibrational *mode*. Figure 3-2 shows six different modes. In mode 1, the entire string vibrates together between the stationary end points. In mode 2, there are two vibrating segments with an additional stationary point in the center. The vibration rate of each mode, called the *frequency* is directly proportional to the number of segments. For

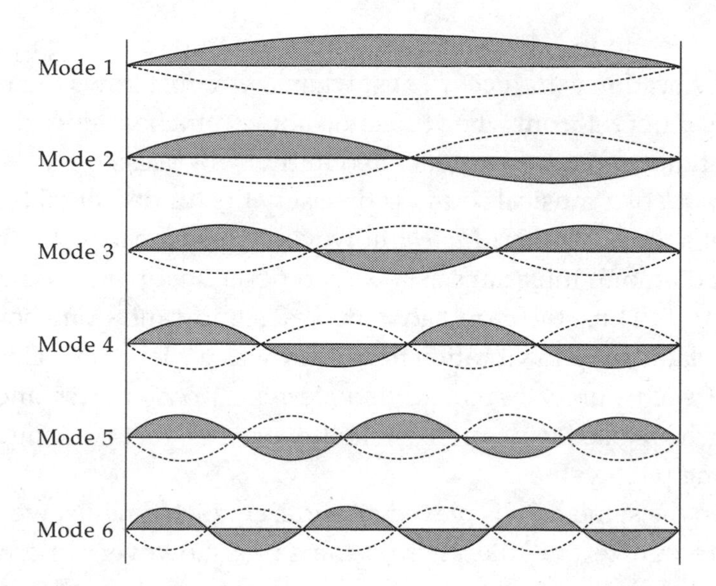

Figure 3-2 Modes of Vibration on a String (From C. A. Taylor, *The Physics of Musical Sounds,* American Elsevier Publishing Co., New York, 1965)

example, in mode 2, with two segments, the string vibrates twice as rapidly (has twice the frequency) as it does in mode 1; and mode 6, with six segments, has six times the frequency. The lowest frequency at which the string will vibrate, represented by mode 1, is called the *fundamental frequency* of the string. The multiples of the fundamental frequency represented by the other modes are called *harmonics, partials,* or *overtones.* The first overtone, or the second harmonic or partial (mode 2), has twice the frequency of the fundamental; the second overtone, or third harmonic or partial (mode 3), has three times the frequency of the fundamental, etc. To minimize the confusion that results from using several notations, from now on I will use the following mixed notation that I feel is the most intuitive: the fundamental is usually the *fundamental,* but occasionally the *first harmonic;* the other modes are the *second, third,* etc. *harmonics.*

When a stimulus is applied to a string, it will vibrate in a combination of these modes, the particular combination depending upon the way the vibration is induced. Thus, when a violinist bows the note A, it sounds different from when the same note is plucked, since different combinations of modes are generated in the two cases.

The *pitch* of a musical sound is precisely the fundamental frequency of the sound. The term *pitch*, however, is psychoacoustic, having originated among musicians as a way to describe a perceived quality of the sound. Thus, the term had a musical significance long before its precise relationship to vibration frequency was understood. This is the origin of such purely psychoacoustic terms as *perfect pitch* and *tone-deaf*, which are used to describe the ability of a person to distinguish or not among pitch values.

When the same note is played on different instruments, we immediately recognize that the pitch is the same, however different the sounds may be in other respects. This property allows the various instruments of the orchestra to tune up. But at the same time, we are aware of the differences, because each instrument generates a unique pattern of harmonics along with the common fundamental. The particular tonal characteristics of the instrument come from the combination of frequencies that result from the specific vibrations of the instrument.

For example, A above middle C is characterized by a pitch, or fundamental frequency, of 440 vibrations per second. Any instrument sounding this note will have one mode vibrating 440 times per second. At the same time, a number of other modes with their characteristic frequencies will be present, the particular combination depending upon the instrument and the way it is played. Figure 3-3 shows some examples of this phenomenon. These pictures are called *waveforms* and show how the amplitude of the acoustical signal varies with time. Such a sound waveform can be obtained from a cathode-ray tube display of the electrical signal derived from a microphone picking up the acoustical signal.

Waveforms *a* and *b* in Figure 3-3 were produced by a violin and an oboe respectively. They look quite different from each other because of

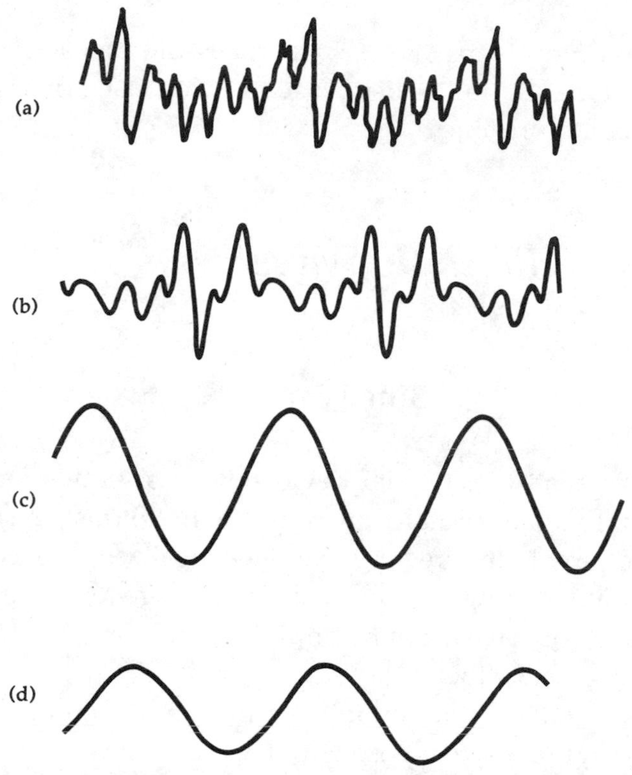

Figure 3-3 Waveforms of A above Middle C (From C. A. Taylor, *The Physics of Musical Sounds*, American Elsevier Publishing Co., New York, 1965)

the different harmonic combinations produced by the two instruments. Nevertheless both waveforms share one common element: each exhibits a pattern that repeats at the same interval, because both have the same fundamental frequency.

The waveform labeled *c* in Figure 3-3 is what is called a *pure tone*. Although it is not produced by a musical instrument, it is very important because it is the waveform that would result if a musical instrument were to vibrate in a single mode with a single frequency. Waveform *d* comes from a tuning fork, and it looks very similar to a pure tone. It is, in fact, about as close as one can get to a pure tone in a musical vibrator. Sounds produced by musical instruments that are rich in harmonics are sums of these pure tones, one for each harmonic. To understand the nature of musical sounds, we must first understand something about pure tones.

Periodic Waveforms

Sine Waves

Musical instruments are complex devices. Because of this complexity, the vibrations that they produce are complex. In contrast, the vibrators that produce pure tones are very simple—idealized, in fact. For example, the simple tuning fork approximates a pure tone. We can gain some insight into just what a pure tone is by examining the elementary vibrators that generate them.

The most fundamental vibrator is a mass at the end of a spring, as shown in Figure 3-4. Suppose you displaced the mass slightly to compress the spring. Being a spring, it tries to decompress and return to its original position. It overshoots that position and, because it is now stretched, it tries once again to reach its original position. Again, it overshoots and becomes compressed. If there were no friction, these successive compressions and expansions would continue indefinitely. If you can imagine a piece of pressure-sensitive paper under the mass moving with a constant speed, the position of the mass would trace out a periodic waveform, and this waveform would have the form of the pure tone shown in Figure 3-3. The frequency of the tone depends upon the size of the mass and the stiffness of the spring: the stiffer the spring and the larger the mass, the higher the frequency. The characteristics of the vibrating string that determine its rate of vibration are ultimately derivable from these mass and spring characteristics. The motion of this elementary vibrator is called *simple harmonic motion*, and the pure

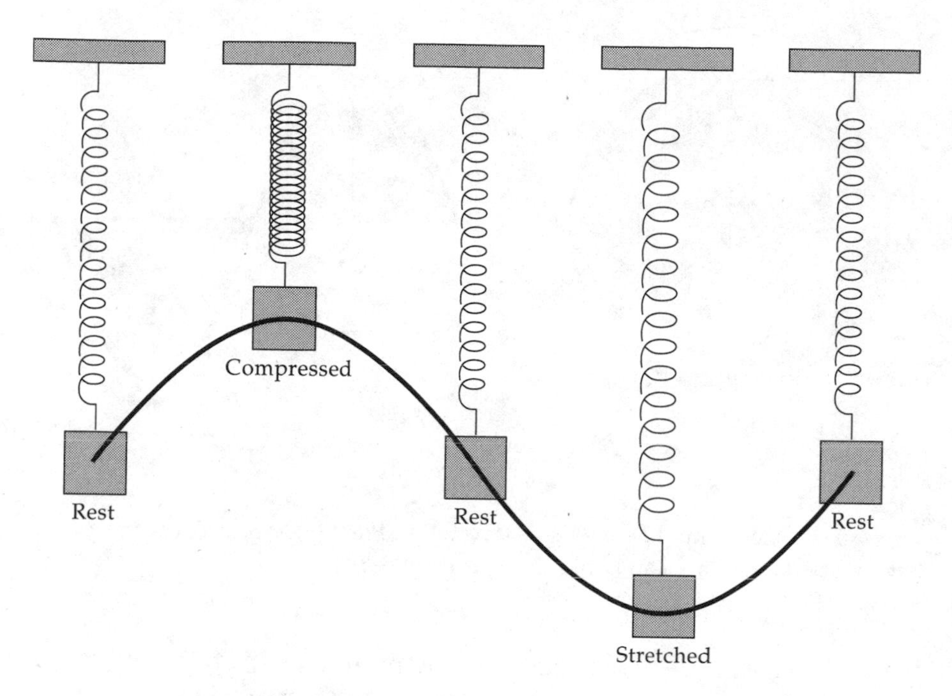

Figure 3-4 An Elementary Vibrator

tone waveform is called a *sine wave*. The source of this name will become apparent presently.

Simple harmonic motion occurs in many idealized situations. It occurs in its purest form only when the deviation of the elementary vibrator from its rest position is slight. For example, a pendulum displaced slightly from the vertical vibrates in simple harmonic motion. If the spring were highly compressed or if the pendulum were displaced from the vertical by a large angle, the motion would deviate from the ideal. A tuning fork is a case in point. The prongs of the fork behave like pendulums. When the fork is struck a slight blow, its motion is about as close to simple harmonic as any musical vibration is apt to get, as indicated by its similarity to the sine wave, as shown in Figure 3-3.

We can show the precise shape of a pure tone or sine wave with the

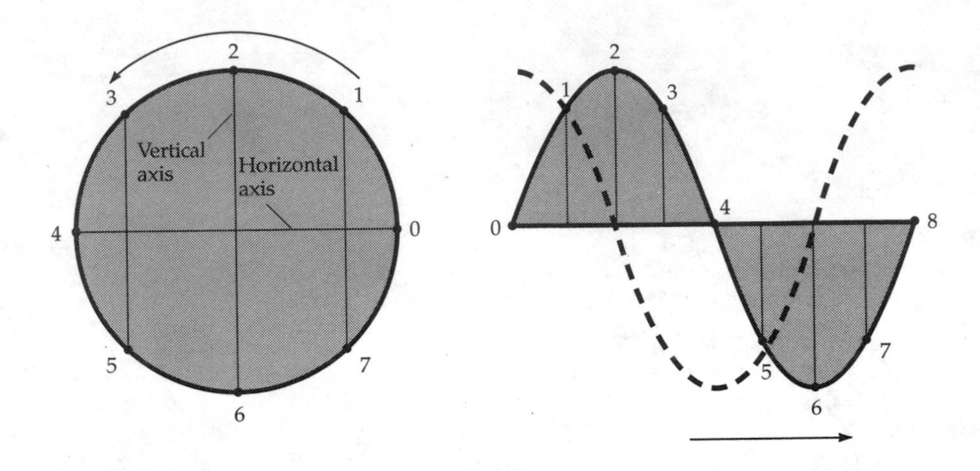

Figure 3-5 Generating a Sine Wave (From J. R. Pierce, *The Science of Musical Sound*, Scientific American Books, New York, 1983)

help of the simple construction shown in Figure 3-5. Let's trace counterclockwise around the circumference of a circle whose radius is one unit in length, at a constant rate of speed, beginning at the point labeled 0. In the figure, it takes 8 seconds to make a complete circuit, and we have marked the position of the trace at each second. Next to the circle, we have laid out a horizontal time axis, analogous to the time axis in the waveform pictures, running from 0 to 8 seconds. At every instant of time, we lay off the height of the rotating point above the horizontal time axis. The result is a curve that looks just like the pure tone of Figure 3-3 and the trace of the mass on the spring in Figure 3-4. If we continue the process, we simply retrace the old path along the circle, and the sine waveform repeats itself. This is what we mean by a periodic waveform.

Why is this curve called a sine wave? The sine wave is related to the *sine function* that you may have first encountered in the study of trigonometry. All the trigonometric functions are properties of angles and are defined by the ratios of the various sides of right triangles. The sine of an angle is defined as the ratio of the leg of a right triangle opposite that angle to the hypotenuse of the triangle. Since the circle shown in Figure 3-5 has a radius one unit long, the height of any point on the circle is the sine of the angle between the radius to that point and the horizontal axis. Therefore the periodic curve in the figure is the sine of that angle plotted as a function of time.

The sine of a 0-degree angle is 0. As the angle increases, the sine increases, reaching a maximum value of 1 at 90 degrees. Then it decreases to a minimum value of –1 at 270 degrees, passing through 0 at 180 degrees.

The *period* of the sine wave is the time it takes for the point to go around the entire circle; in our example, 8 seconds. This corresponds to the time interval between corresponding points of successive repetitions. The frequency of the sine wave is the reciprocal of the period; that is, the number of complete rotations made in one second. In our case, it is 1/8 of a revolution or cycle per second. Until recent years, the unit used for frequency was called the *cycle per second*. One cycle per second is now called a *hertz* (abbreviated Hz), named after the nineteenth century German physicist, Heinrich Hertz. As we saw before, the frequency of the fundamental tone or pitch of A above middle C is 440 Hz, and its period is 1/440th of a second, or 0.00227 second.

Phase

The cosine of an angle is defined as the ratio of the leg of a right triangle adjacent to the angle divided by the hypotenuse. In the picture shown in Figure 3-5, the cosine is the distance from a point on the circle to the vertical axis. Its trace with time is shown by the dashed curve in Figure 3-5. The cosine has exactly the same form as the sine (the term used is sinusoidal), but it starts at 1 instead of at 0. It then drops to 0 at an angle of 90 degrees and continues on in the same way.

The *phase* of a sinusoidal wave designates the angular difference between the its starting point and that of a sine wave itself.

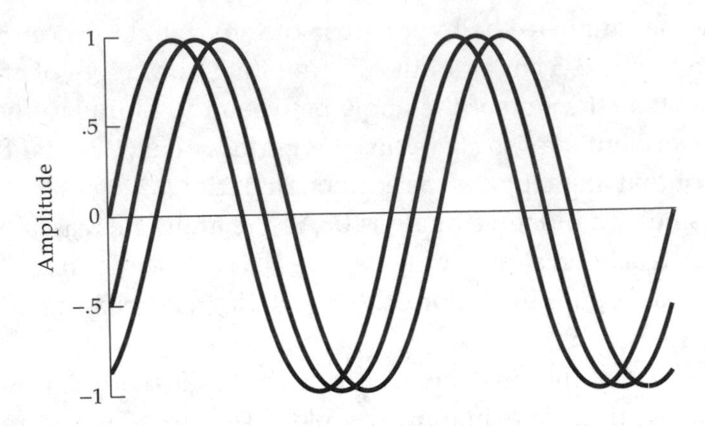

Figure 3-6 Sinusoids with Different Phases

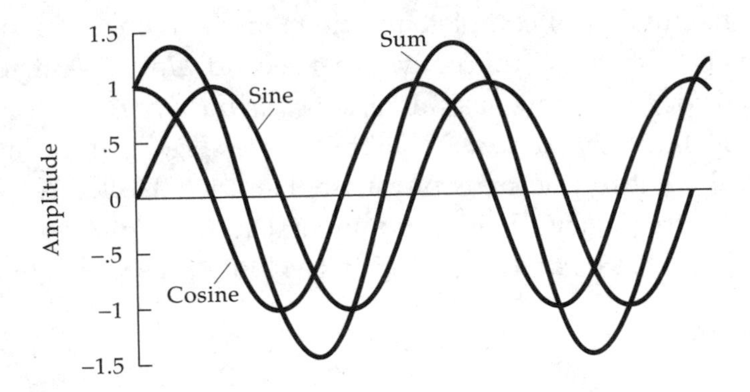

Figure 3-7 When You Add Sinusoids, You Get Sinusoids

You can see from Figure 3-5 that the cosine wave has a phase of 90 degrees with respect to the sine wave. Figure 3-6 shows sinusoids with different phases. If you add two sinusoids of the same frequency, but different amplitudes and phases, the result is another sinusoid of the same frequency, with the phase intermediate between those of the two original sinusoids and the amplitude somewhere between the sum and difference of the two amplitudes. As an example, Figure 3-7 shows the result of adding a sine and cosine with the same frequency and amplitude. The sum is a sinusoid with a phase of 45 degrees and an amplitude 1.4 times the amplitudes of the component sine and cosine.

Musical Waveforms

These elementary properties of sinusoidal waveforms provide the foundation for understanding more about the nature of complex musical waveforms. Recall that any musical sound is a sum of harmonics derived from its modal vibrations. This is equivalent to saying that a musical sound is the sum of a number of sinusoids. An example should make this clear. Trace *a* of Figure 3-8 repeats the oboe waveform of Figure 3-3. The other traces show a sequence of approximations to the oboe waveform, each closer than the previous. In each approximation, some number of harmonics are added, each with a particular amplitude and phase. Trace *b* sums up the first three harmonics. Even with three harmonics, the gross shape of the waveform begins to look like the original. With the addition of two more harmonics in trace *c*, some of the fine structure of the original begins to show up. The addition of the sixth and seventh harmonics in trace *d* gives fairly respectable agreement with the original. Finally, the ninth and tenth harmonics are added in to form trace *e*, which is even closer to the original.

All the synthetic waveforms shown in Figure 3-8 are constructed of particular sine waves with particular amplitudes and phases. Had we chosen other values for these parameters, the resulting waveform shapes would have been quite different. How do you find out what the correct parameters are? One way is to do an exact analysis of the vibrations of the air in an oboe, but that's extremely hard to do. Another way is to use a trial and error technique: start out with the fundamental and add in a little second harmonic, then try a little third, etc., similar to the way a chef seasons a sauce. That doesn't work very

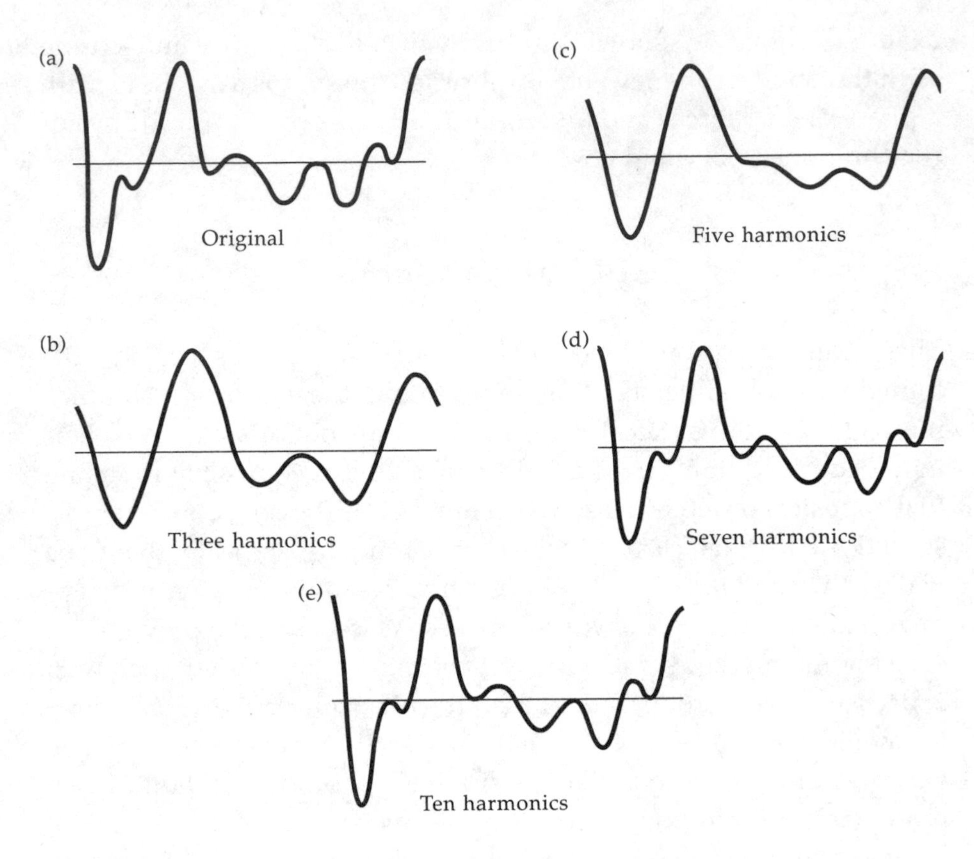

Figure 3-8 Reconstruction of a Waveform from Its Harmonics

well either, because there are too many combinations of amplitudes and phases. A 10 degree phase difference or a 10% amplitude difference is much more significant than 10% too much oregano or thyme. A more successful approach is to use a very powerful mathematical technique called *Fourier analysis* which gives a precise formula that allows us to compute the amplitudes and phases of all the harmonics from knowledge of the waveform. This technique was used to generate the table on the facing page, which lists harmonic amplitudes and phases for the oboe waveform of Figure 3-8a.

Harmonic	*Amplitude*	*Phase, degrees*
1	1	0
2	1.8	174
3	2.1	126
4	1.1	92
5	0.2	181
6	1.5	97
7	1.1	75
8	0.3	86
9	0.4	56
10	0.3	–6

The fundamental, a sine wave of amplitude 1 and phase 0, serves as a reference for all the others. The largest harmonic is the third and the smallest is the fifth. The eighth, ninth and tenth are relatively small, which is why trace *e* in Figure 3-8 is not very different from trace *d*.

The method for calculating the amplitudes and phases of the harmonics of a periodic waveform from the waveform itself is due to the celebrated nineteenth century French mathematician, Jean- Baptiste Joseph Fourier. Any periodic waveform, regardless of the physical mechanism by which it is generated, can be represented as a sum of harmonics in this way. It is easy to appreciate why this is true for musical instruments, since the modal harmonics result directly from their vibrating strings or columns of air. But the power of Fourier's method lies in its generality, which has made it enormously important to many fields.

Musical Harmonics

In the last example, we saw that 10 harmonics gave an excellent approximation of the oboe waveform. If we add more harmonics, the approximation improves even more. But the comparison of waveforms is a means to an end, not an end in itself. The important

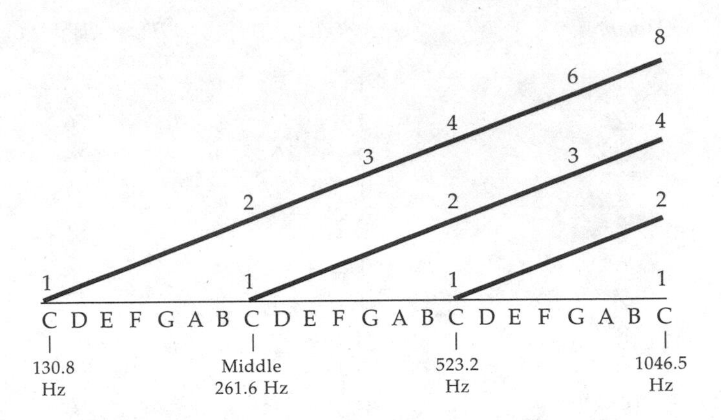

Figure 3-9 Harmonic Relationships

question is, how many harmonics are needed to reproduce the sound of the oboe note? This is very hard to pinpoint quantitatively—the ultimate answer is in the ears of the listener. Certainly, the waveform comparison is a guide. But the relationship between this guide and our ears is not straightforward. The closer the waveforms, the closer the sounds must be, but it is difficult to associate the closeness of the waveforms with the degradation of the approximation to the original.

A more direct way to judge whether a particular harmonic is necessary is to determine whether its effects are audible. You can do experiments of this kind at the piano. As noted earlier, some of the characteristic sounds of a piano arise from the excitation of strings caused by the vibration of other strings. The experiment that will be outlined gives an indication of this effect.

The chart in Figure 3-9 represents the harmonic structure for three octaves of a piano keyboard beginning at C below middle C. When any of these notes is sounded, a set of harmonics is produced, in addition to the fundamental. The harmonics produced are shown on the diagonals in Figure 3-9. For example, since a musical octave represents a factor of two in frequency, the second harmonic of C below middle C occurs at the same frequency as the middle C fundamental. Similarly, the third harmonic has approximately the same frequency as the fundamental of G above middle C. The fourth harmonic of C below middle C is the second harmonic of middle C and the first harmonic (fundamental) of C above middle C.

The piano strings are normally damped so that they cannot vibrate. Depressing a key causes a hammer to strike the string and simultaneously lifts the damper, thus allowing the string to vibrate freely. Our experiment is to press C above middle C gently. This prevents the note from sounding, but lifts the damper from the string so that it is free to vibrate. Now strike middle C sharply. The string corresponding to middle C vibrates only while its key is depressed; once the finger is lifted off the key, the damper stops further vibration. But you continue to hear a relatively faint lingering tone one octave above middle C. Why? Because the second harmonic component of the sound excites the undamped string of the note one octave higher at its fundamental frequency. Now do the same experiment in reverse. Depress middle C without sounding the note and strike C above middle C sharply. This time you hear the tone of C above middle C (about 520 Hz) coming from middle C. This tone is caused by the middle C string being excited at its second harmonic. These sympathetic vibrations are used by piano tuners to adjust the upper and lower registers of a piano after adjusting the middle range with the aid of tuning forks. You can continue the experiment using higher harmonics to see how many are actually audible.

Figure 3-10 shows the range of the fundamental frequencies for the entire piano keyboard. For comparison, the ranges of many other musical instruments are also shown. Note the enormous range of the piano—it extends from 27.5 Hz on the low end to 4186 Hz on the high end, a band of more than 4000 Hz (or 4 kilohertz, abbreviated kHz). Because each of these fundamental frequencies is accompanied by a number of harmonics when the note is sounded, the range of frequencies produced by the piano can extend well beyond 20 kHz, since this frequency is between the third and fourth harmonic of the top note on the keyboard. It is also important to note that 20 kHz is the highest frequency usually perceptible to a person with normal hearing.

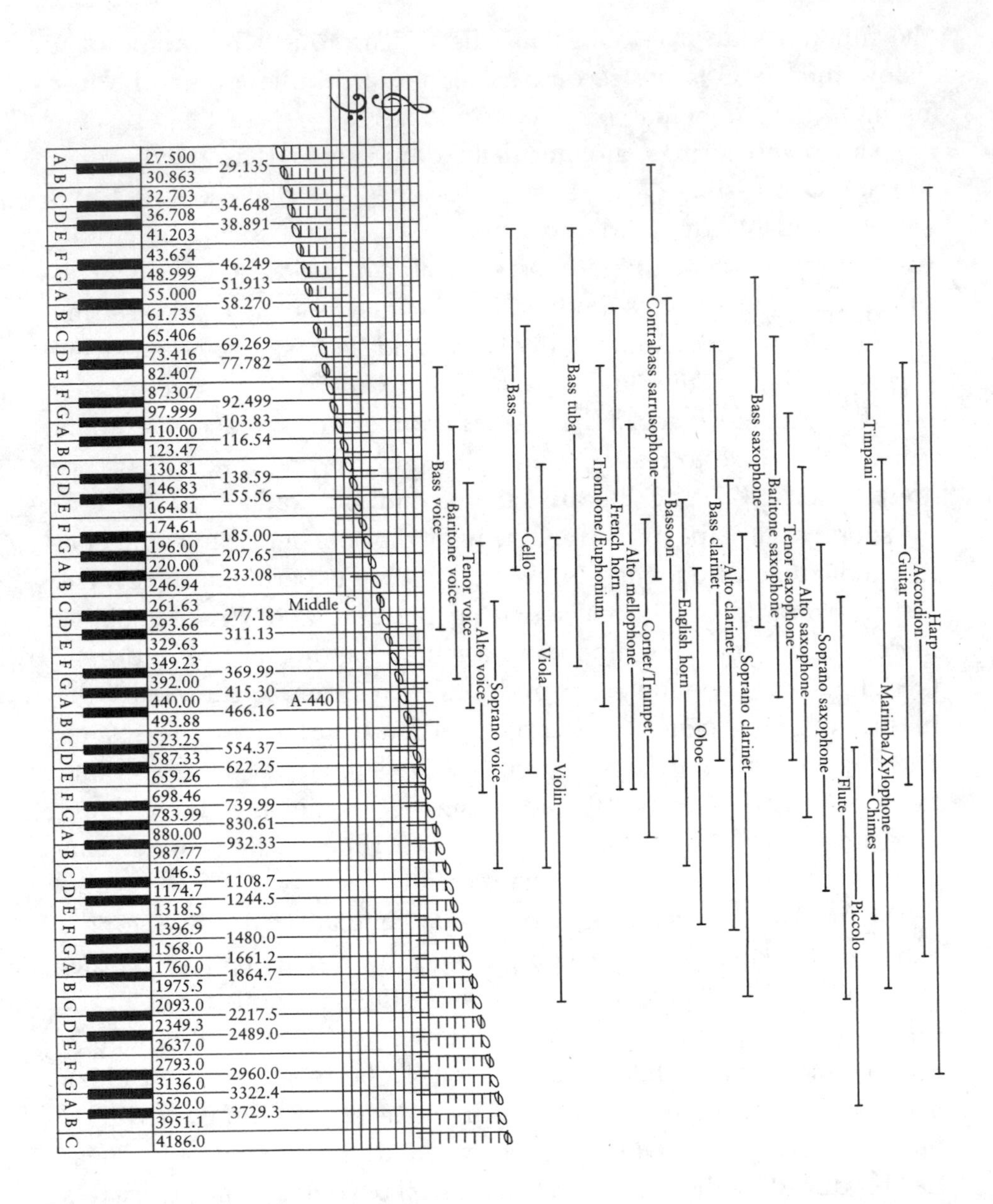

Figure 3-10 Pitch Ranges of Musical Instruments (From J. R. Pierce, *The Science of Musical Sound*, Scientific American Books, New York, 1983)

Speech

Speech Signals

The human voice is a musical instrument, indeed, the most extraordinary of all. Figure 3-10 also includes the frequency ranges of the singing voices among the other instruments. The combined ranges of the five defined singing voices come close to spanning the ranges of the cello and viola. Whether used for talking, singing, or shouting, the voice is indeed an instrument with all the attributes of other instruments and then some.

Since the ultimate origin of all sound is a physical vibration, the human voice, too, derives its properties from physical vibrations. The sources of these vibrations are in our bodies: the vocal cords, all the passages of the head, and the lips, tongue, teeth, etc. Since we can exercise a great deal of control over these organs, we can achieve very great variation in the way we use our voices in speaking and singing. We may marvel at the expressiveness derived by a great musician from a violin or piano. But the range of variations in playing these instruments, great as it may be, is eclipsed by the range of expressiveness that a great singer can command.

The Physiology of Speech

People can generate many different kinds of sounds. Even the simplest categorization into vowels and consonants can tell us a great deal. Compare the sound "ah" with the sound "sss," for example. The "ah" can be spoken or sung in a large variety of ways. If sung, we can sound it at any pitch that we can produce. If you place your finger on your Adam's apple while sounding the "ah," you can feel that the vocal cords are vibrating. In contrast, the hissing sound "sss" has very few variational possibilities. To make such a sound, you simply force air through your teeth, and the sound results from the turbulent air flow. There is no recognizable vibrational mechanism taking place in the vocal cords or anywhere else.

In this example, the "ah" represents a vowel, and the "sss" is called an *unvoiced consonant*. Pay attention to what you do when you say "zzz" instead of "sss." You hold your facial muscles in just about the same position for the two consonants. The difference is that when you say "zzz," you are sounding your voice rather than simply hissing. You can vary the pitch as when pronouncing a vowel, and you can feel the vibration on your Adam's apple. "Zzz" is an example of a *voiced consonant*.

The drawing in Figure 3-11 indicates some of the physiological basis for the human voice-producing mechanisms. All speech sounds begin with the lungs as the energy source. Air is forced from the lungs through the various anatomical passages and ultimately finds its way out of the mouth. The source of voiced sounds, vowels, or voiced consonants, is the vocal cords. Air from the lungs passes through these cords, which vibrate back and forth, opening and closing the space in between, the glottis. This produces periodic puffs of air at the vibration rate of the vocal cords. We can vary the vibration rate of the vocal cords by varying the tension in the muscles comprising the chords. Figure 3-12 shows an expanded view of the vocal cords along with the thyroid cartilage, which produces the protrusion from the neck that we call the Adam's apple.

The pitch of the voice is the fundamental frequency of the vocal cord vibration. The pitch range for male voices when speaking is generally between 80 and 240 Hz. Since women have shorter vocal cords than men, their pitch is somewhat higher, ranging between 140 and 500 Hz for speech. The pitch range for singing extends well beyond that for speech. A good tenor can sing a high C (C above middle C, or 523 Hz), while a good soprano can sing an octave higher. And like the vibrators that we have seen in musical instruments, vibrations of the vocal cords are rich in harmonics.

The vocal cords play no role in the production of unvoiced sounds.

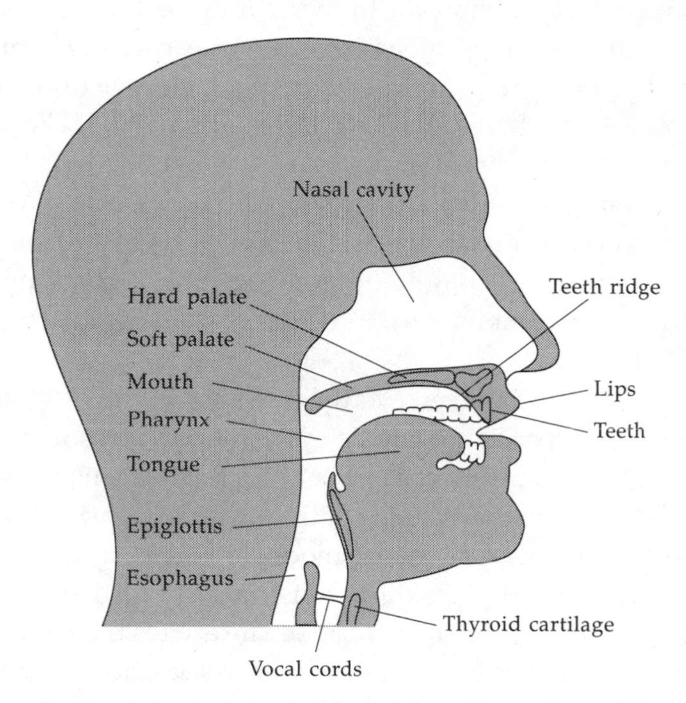

Figure 3-11 The Physiology of Human Speech Production (From W. J. Strong and G. R. Plitnik, *Music, Speech and High Fidelity,* Brigham Young University Press, Provo, Utah, 1977)

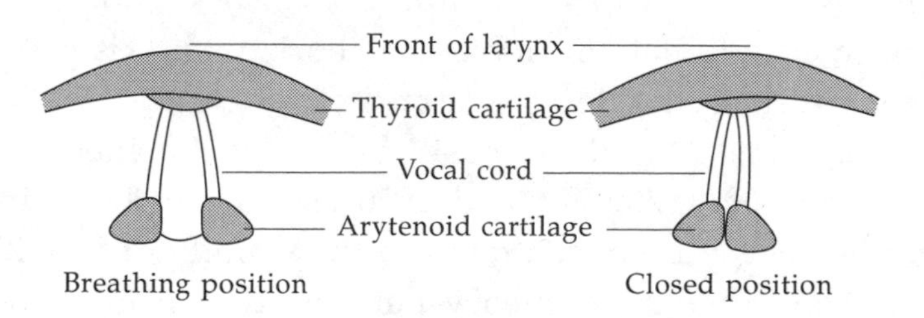

Figure 3-12 A Simplified Top View of the Vocal Cords (From W. J. Strong and G. R. Plitnik, *Music, Speech and High Fidelity,* Brigham Young University Press, Provo, Utah, 1977)

To produce these sounds, air from the lungs is forced into the head through other orifices, bypassing the vocal cords. We produce an unvoiced consonantal sound by creating a constriction somewhere in the mouth with the teeth, lips, or soft palate. The air forced through such a constriction becomes turbulent or noise-like, creating the hissing sounds associated with such sounds. Voiced consonants combine the two effects: periodic puffs of air from the vocal cords are passed through the constrictions in the mouth, thus generating the combined sound patterns characteristic of voiced consonants.

Speech acousticians refer to the passages in the head and mouth that shape the speech sounds as the *vocal tract*. In some sense, it is analogous to the tube in a clarinet or oboe. But it is much more flexible. You can vary the effective length of the musical instrument and, hence, the frequency of its vibrational modes by opening and closing the holes along the tube. A skilled musician can also introduce some measure of variation by the way he or she excites the reed with his or her lips. But that's the extent of it. The vocal tract can be varied in many ways, some of which we have already noted. The very large number of vowel sounds are made by holding the mouth, tongue, and lips in different positions. As we do this, we change the frequency modes in various complex ways.

Each vowel or other voiced sound is usually characterized by energy peaks clustered around three or four frequencies known as *formants*. Figure 3-13 shows examples for two sounds, *a* as in *had* and *ee* as in *heed*. In both examples, the horizontal axis is frequency and the vertical axis is energy. The curve in Figure 3-13*b* shows that when the *a* in *had* is spoken, most of the acoustic energy is in the vicinity of 660 Hz, 1700 Hz, and 2400 Hz. When the *ee* in *heed* is spoken, the energy is concentrated in the vicinity of 200, 2200 and 3200 Hz. The different frequency patterns shown in these curves account for the differences that we perceive in the sounds. And, as noted above, they result from the different ways in which we hold our facial muscles when making the two sounds.

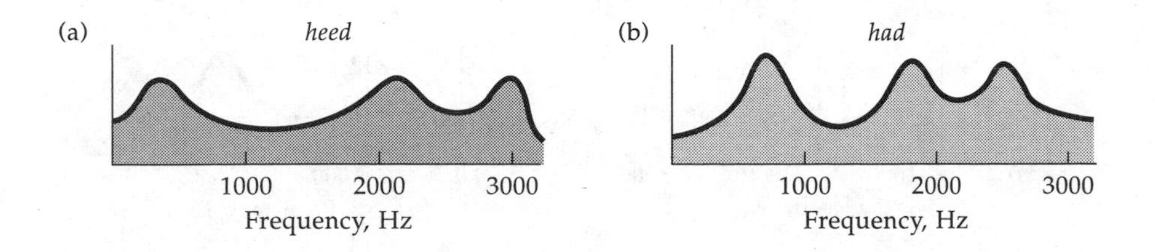

Figure 3-13 Formant Positions for Two Vowels (From J. R. Pierce, *The Science of Musical Sound,* Scientific American Books, New York, 1983)

Speech Frequencies and Harmonics

The formant curves in Figure 3-13 show the relative amounts of energy in the various frequency components of the speech sounds. Such curves are examples of *frequency domain representations.* The waveforms that we have been using up until now show how the energy in music and speech sounds varies with time, and for this reason they are sometimes referred to as *time-domain representations.* Both representations are useful and we will see more of them. But for now, we can use the frequency domain representation to tell us more about the characteristics of speech.

A person produces a desired sound by tensing or relaxing the vocal cords to vary the pitch, as well as by varying the vocal tract. Therefore the characteristics of the resulting speech has to reflect both vocal cord and vocal tract phenomena. Figure 3-14a shows how the acoustic energy is distributed over the harmonics of vocal cords sounding a pitch of 200 Hz. The curve indicates that the same amount of energy is contained in each of the harmonics. If we could construct a musical instrument that consisted of an air source similar to the lungs that forced air through a pair of vibrating membranes similar to the vocal cords and then directly into the air, the resulting waveform would be composed of a sum of sine waves at the harmonics of the vibrator, all with the same amplitude.

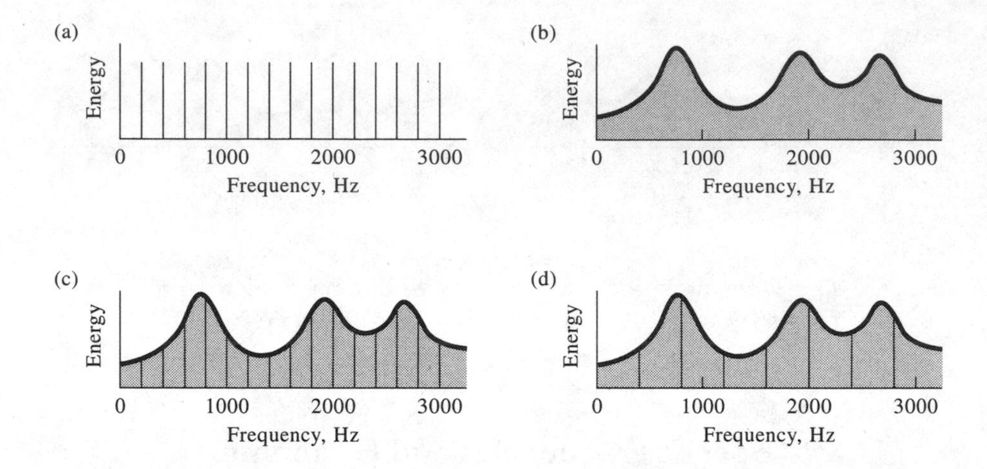

Figure 3-14 Speech Frequency Spectra (From J. R. Pierce, *The Science of Musical Sound*, Scientific American Books, New York, 1983)

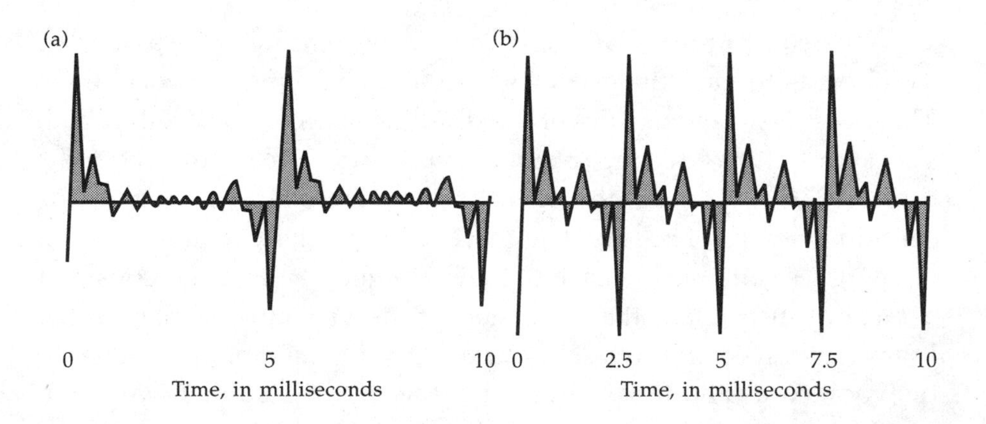

Figure 3-15 Vowel Waveforms

But that sound would not be that of the human voice because it would not reflect the behavior of the vocal tract. In the human voice the frequency components coming from the vocal cords are modified by the formant structure of the vocal tract. When this happens, each of the vocal cord harmonics assumes the amplitude of the vocal tract's formant curve, as is shown in Figure 3-13, repeated for convenience in Figure 3-14b. Figure 3-14c shows the result when a man speaking with a pitch of 200 Hz makes the sound of the *a* in *had*. Suppose a female were to sound the same vowel as the male, but with a pitch of 400 Hz instead of the male's 200 Hz. Her vocal cord energy would be concentrated equally at 400 Hz and its harmonics. Thus, her vowel sound energy, shown in Figure 3-14d, would be concentrated at the harmonics of 400 Hz, but with amplitudes modified by the same formant curve of Figure 3-14b.

We can compute the waveforms associated with these *frequency spectra* by adding up sine waves with the amplitudes given by Figure 3-14. When we do this we get the waveforms or time domain picture shown in Figure 3-15. Both the frequency and time-domain pictures give us the same information, but in different forms. Sometimes the one or the other is of more interest to us. Fourier analysis mentioned earlier gives us the method for going back and forth between the two pictures.

A very convenient way to show how speech varies in both time and frequency is called the *sonogram*. The example shown in Figure 3-16 is for the phrase "speech communications." In the figure, time goes from left to right and frequency from bottom to top. The sound energy is represented by the darkness of the trace. Note first the well-defined formant structure for the vowels that is entirely absent in the unvoiced sounds (*s, ch, c, ti*). Particularly significant is the fact that most of the energy in the voiced sounds is below 4 kHz, while most of the energy in the unvoiced sounds is above 4 kHz.

The consonants convey most of the information content of speech. You can prove this to yourself by writing down some text, omitting the vowels, and asking someone to read it. The person should be able to

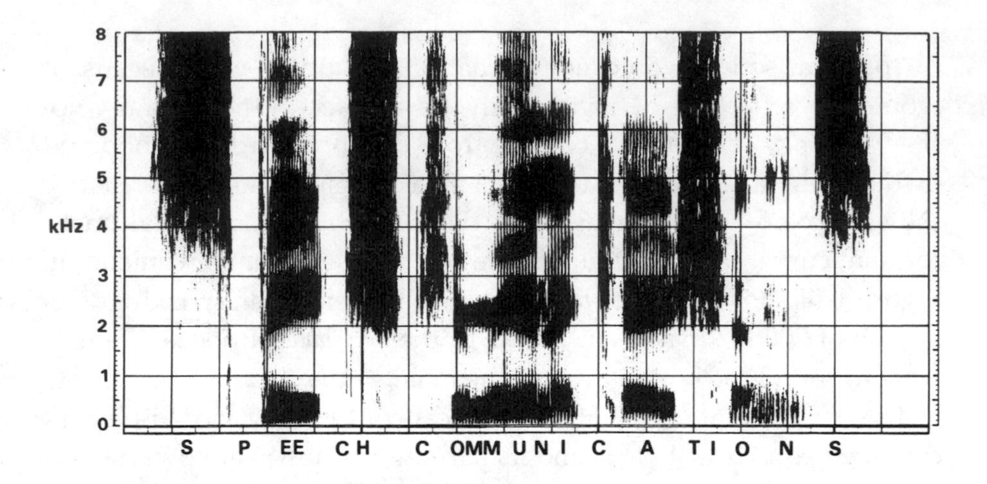

Figure 3-16 A Speech Sonogram (From The Massachusetts Institute of Technology, Lincoln Laboratory)

do so quite easily. If you do the reverse, omit the consonants and leave in the vowels, the person will not be able to understand anything at all. Understanding speech is quite analogous. The intelligibility of speech depends to a large extent upon how well the consonants are articulated. Since so much of the energy content of the unvoiced consonants lies at the higher frequencies, those frequencies must be present for high-quality, high-intelligibility speech.

Bandwidth

To communicate any piano note—say middle A—faithfully, all the harmonics produced when that note is struck must be transferred from the source (a piano) to a destination (a listener's ears). If any harmonics are missing, the sound heard by the listener will be different from the sound emitted by the piano. How different it will sound is very complicated, since our perception of the sound is a very complex function of the actual acoustic signal. Nevertheless, it is safe to say that a modification of the signal will most likely modify the sound as perceived by the listener. If the piano and the listener are in the same room, there should be no problem; the channel in this case has very high fidelity. But if they are remote from each other, there may well be a problem.

We don't usually listen to music over the telephone, and for good reason. A typical telephone line will pass frequencies between 300 and 3000 Hz. This means that, generally speaking, frequencies in that range will be passed reasonably (but not perfectly) faithfully, and that frequencies outside that band will be attenuated—the farther away, the greater the attenuation. This means that middle A's seventh harmonic at 3080 Hz may be slightly attenuated, its eighth harmonic at 3520 Hz will be highly attenuated, and any higher harmonics will not be transmitted at all. That's probably not too serious; the first seven harmonics should reproduce the note reasonably well. But there are many notes above middle A on the piano. In fact, the A three octaves above middle A, with a fundamental frequency of 3520 Hz—the same as the eighth harmonic of middle A—will suffer the same fate, and there are a few notes above it on the keyboard. The problem is even worse at the low-frequency end. Middle C has a fundamental frequency of 261 Hz, somewhat below the lower edge of the band. Clearly, the telephone channel is a terrible channel for the transmission of music.

But we are accustomed to using the telephone for transmitting speech signals. It is clear from the sonogram in Figure 3-16 that the telephone does not pass most of the energy in the unvoiced consonants (the *sp* and *ch* in *speech* and the *c*, *ti*, and *s* in *communications*). Since these sounds convey much of the intelligibility information in the speech, we see why we miss many sounds on the telephone that cannot be filled in by the brain from the context of the sentence's subject matter. But if you were to spell out a word, it would be hard to distinguish a *t* (*tee*) from a *c* (*see*). However, most of the time telephone lines give us adequate intelligibility for the uses to which we put them.

The term *bandwidth* is a shorthand way to designate the width of a band of frequencies. Thus, the bandwidth of the acoustic signals produced by the piano is in excess of 20 kHz, and the bandwidth of speech signals is around 7 kHz. The amount of information in a music or speech signal is proportional to the bandwidth of the signal. If we want to send speech with high fidelity, the communication system, be it analog or digital, must pass all the audible frequency components comprising the speech. If it does not, information is lost. The design of communications systems always includes a compromise between fidelity or quality and cost. The telephone system represents a compromise between fidelity and economy, based largely upon the analog technology of the past. As we shall see later, modern digital computer and communications technology is rapidly changing the basis for this compromise.

4

Transmitting Audio and Video

The older telecommunications services—telephone, radio, and television—started out as analog systems and have remained so, by and large. Before the development of the computer, it would have been inconceivable for anyone to convert the human voice to a string of digits before transmitting it over a wire or through the air. Even today, when digital communications are commonplace, converting the audio signal to a stream of 0s and 1s and then sending the digits isn't the first thing that comes to mind. The natural thing to do is to transmit the audio as it comes, in analog form.

Talking over the Telephone

The telephone line is the most familiar communications channel. It is also the simplest conceptually. Your local phone company provides a connection between your telephone and its *end office* in the form of a pair of wires. When you talk into the microphone built into the telephone handset, your voice signal is converted from acoustical to electrical form and is carried in this way to the end office. From there it goes through an elaborate network of transmission and switching facilities (largely digital) until it reaches another pair of wires leading to the telephone of the person to whom you are talking. Despite the complexity of this network, we can think of the telephone line as a long pair of wires connecting the two telephones.

This circuit delivers a distorted version of your voice to its destination. We can gain some insight into the nature of this distortion by comparing telephone communication with some other higher fidelity audio systems. Figure 4-1 shows a succession of audio communication channels of decreasing fidelity. The top picture (*a*) shows the best of all possible channels, the standard against which every other channel must be compared: the pure acoustical channel with no intrusion of electrical signals at all. The next picture (*b*) shows the best of all possible electrical channels: a high-quality microphone, a short pair of wires and a pair of high-fidelity earphones. The microphone and earphones represent the best achievable energy converters (*transducer* is the general term), and the wires connecting them are short enough that they do not introduce degradation. The sound quality that results from this channel is very good, but still is not quite as good as that of the purely acoustical channel, because even the best microphone and speaker introduce some degradation. In the next picture (*c*), the high-quality microphone and earphones are replaced by the lower quality devices found in the ordinary telephone handset; but, as before, the handsets are connected with a short pair of wires. The telephone transducers themselves are significantly poorer than the high-fidelity

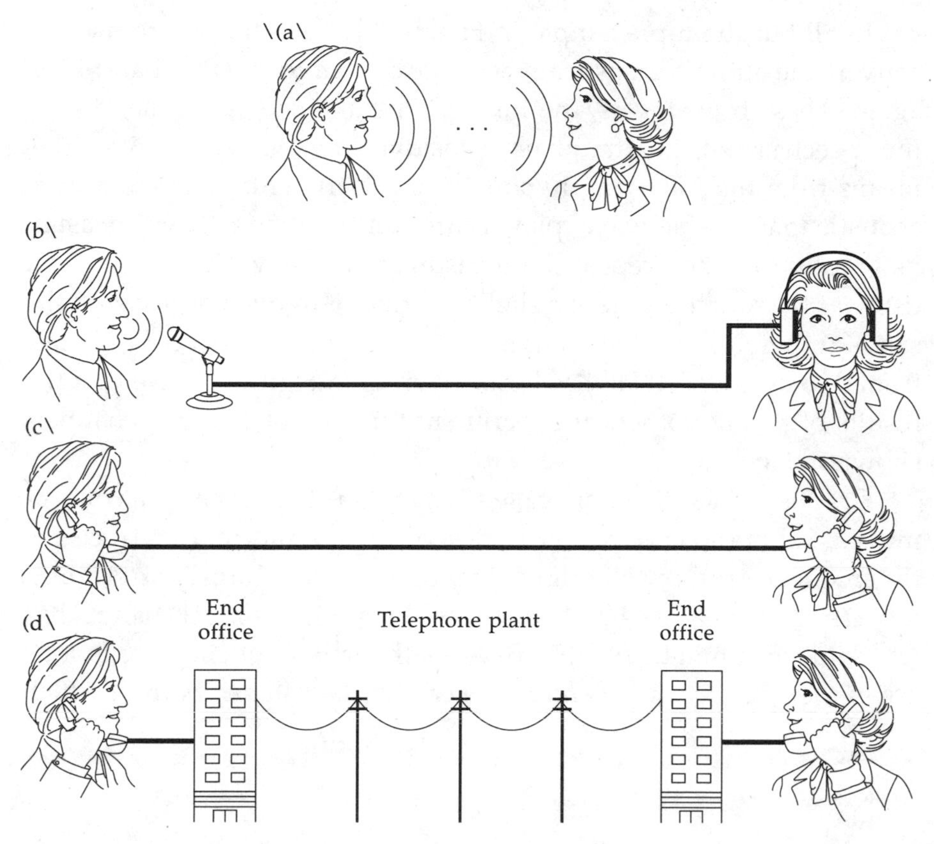

Figure 4-1 Audio Channels

equivalents in the previous example. Thus, even though the circuit connecting them does not introduce any degradation, the resulting audio quality is significantly lower than it was in case (*b*). The final picture (*d*) shows a telephone circuit with telephone handsets connected by a realistically long pair of wires. As you would expect, the resulting quality is still worse, because of the circuit degradations.

The physical properties of the telephone channel that influence its capacity and, hence, degrade the quality are signal-to-noise ratio and bandwidth. Having addressed these two factors individually in the last two chapters, we now examine their combined effects.

In all but the top example in Figure 4-1, transducers are used to convert the audio signals from acoustical form to electrical and back again. These transducers, the microphone and the earphone, distort the speech signals in a frequency-sensitive way. We can see what this means from the following hypothetical experiment: generate a pure acoustic tone or sine wave, play it through the channel, and measure its amplitude. Then repeat the measurement many times, each time using as the source a tone at a slightly different frequency, but with the same amplitude, until the entire audible frequency band of signals from 0 to over 20,000 Hz has been covered. In Figure 4-2 we plot the results of this hypothetical experiment for each of the four channels shown in Figure 4-1.

Graphs such as these are called *frequency-response curves*, a concept introduced in the last chapter to describe speech formants. Trace *a* is the frequency response of the reference acoustic channel corresponding to trace *a* in Figure 4-1. As one would expect, this ideal channel has no frequency limitations whatsoever—the only frequency limitations are in the musical source and in the ears of the listener, both of which

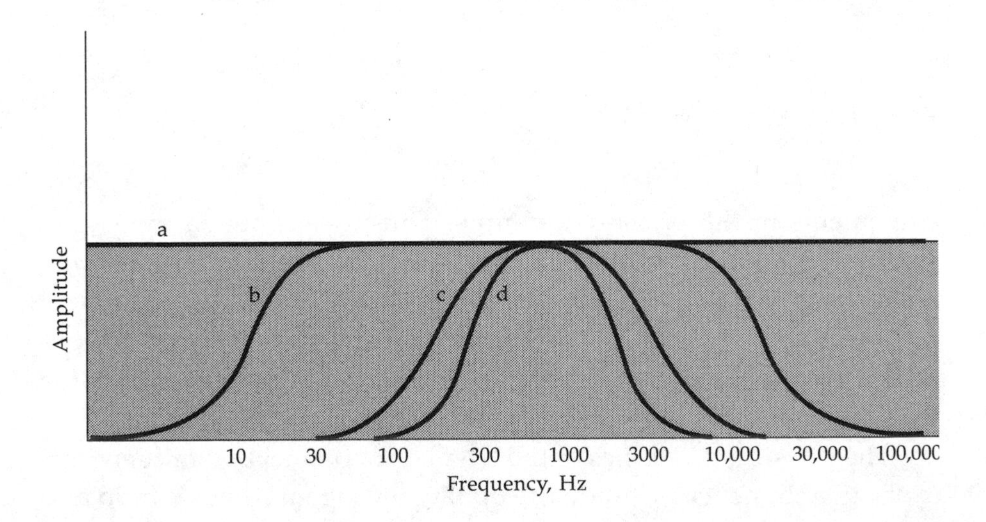

Figure 4-2 Frequency Response of Audio Channels

are external to the channel itself. Traces *b* and *c* show the frequency-response characteristics of the corresponding channels of Figure 4-1. The differences are readily apparent and are due to the effects of the different microphones and speakers. In both cases, the channel passes a particular band of frequencies and rejects all others. The high-fidelity channel in trace *b* has a frequency response that passes a wide band between 30 and 20,000 Hz. The lower fidelity channel in trace c passes the much narrower band between 250 and 3500 Hz. Or, in terms of bandwidth, the high-fidelity channel has a bandwidth of 19,970 Hz, and the lower fidelity channel has a bandwidth of 3250 Hz. When you buy a microphone or a loudspeaker, the manufacturer usually provides a frequency-response curve similar to the curves shown in Figure 4-2. These curves describe the performance of the item.

Any device that exhibits the characteristics of these channels—i.e., that treats the different frequencies selectively—is called a *filter*. All our examples have been of *bandpass filters*, which pass a band of frequencies within a total frequency band of interest and reject frequencies above and below that band. There are other common varieties of filters. For example, a *lowpass filter* passes all frequencies below some threshold frequency, and a *highpass filter* passes all frequencies above a threshold frequency. Ordinary walls are lowpass acoustic filters, since they attenuate treble tones much more severely than they do bass tones.

A real telephone channel has a frequency-response characteristic that is even narrower than that shown in Figure 4-2 for the lower fidelity channel (trace *d*), due to the effect of the telephone connection itself. A telephone line distorts the audio band a great deal. Every once in a while when you call some business establishment and are put "on hold," you are forced to listen to music through the telephone system, under the misguided notion that the music will soothe you while you wait. If you have been subjected to this, you know that the quality of the music is poor, at best, and downright unpleasant, at worst. This is because the telephone system was not designed to transmit music. Its bandwidth of around 3000 Hz is considerably narrower than that needed for most kinds of music. The telephone system was designed to carry the human voice, and it does a reasonably good job. But even in this case, the bandwidth is narrower than what is needed for complete intelligibility, since many of the consonant sounds in speech have significant energy above the cutoff frequency of telephone circuits.

The other effect introduced by a real telephone line is noise. Many phenomena in the complex telephone plant introduce noise into the circuit. A sine wave, transmitted within the bandwidth of the circuit, is received at the other end with random variations superimposed upon it. Some of these variations are small and some are large. The audible effect is the presence of extraneous sounds that, if large enough, mask speech either partly or totally. Large impulses sound like "pops." Continuous low-level thermal noise sounds "hiss"-like. When the noise level is low, it doesn't interfere with speech intelligibility; the primary limitation is the bandwidth. But when the noise level is high enough, its effects combined with those of the frequency limitations to make some or all of the speech unintelligible. The words "high" and "low" make sense only when related to the signal amplitude. A voice signal loses strength as it travels over a long phone line. The telephone company amplifies the signal to compensate for this loss. The resulting signal, as it appears in your earphone, is usually in the correct loudness range for your ears. But whether you can understand what is spoken is determined, in part, by the signal-to-noise ratio.

Radio Broadcasting

The long-distance portion of the telephone plant uses a variety of media to transport telephone signals. Some are cables, either electrical or optical, and some use radio or *wireless* techniques. Indeed, the evolution of these transmission facilities in the public networks can be characterized as the search for more economical sources of the bandwidth required to satisfy the ever- increasing telecommunications demand. But since the ordinary analog circuit with a bandwidth of 3 kHz is the commodity that the subscriber sees, the fact that wide-band media exist is largely invisible. They are simply vehicles for carrying bundles of these narrowband circuits over long distances. The broadcast media are another matter entirely. The various radio and television channels have wider bandwidths that are evident to all. While many of the characteristics of these broadcast channels are similar to those of the telephone circuit, there are some significant differences.

The steps involved in transmitting audio by radio are shown in Figure 4-3. Note that we can't simply talk into a radio transmitter as we can into a telephone or tape recorder, both of which are designed to receive audio signals directly. They accept signals in the audio frequency band. In contrast, before an audio signal can be broadcast over the radio, it must be moved to a higher frequency range by a process called *modulation*. The two common forms of modulation used in broadcasting, *amplitude modulation* and *frequency modulation*, are familiar to us by their respective abbreviations, *AM* and *FM*. These higher frequency electrical signals are fed to an antenna that converts the signals to *electromagnetic waves* that are radiated out into space. When these waves are received by another antenna, they are converted back to electrical signals. The audio waveform is then extracted from the high-frequency electrical signals by the inverse process, called *demodulation*, and then played for the listener.

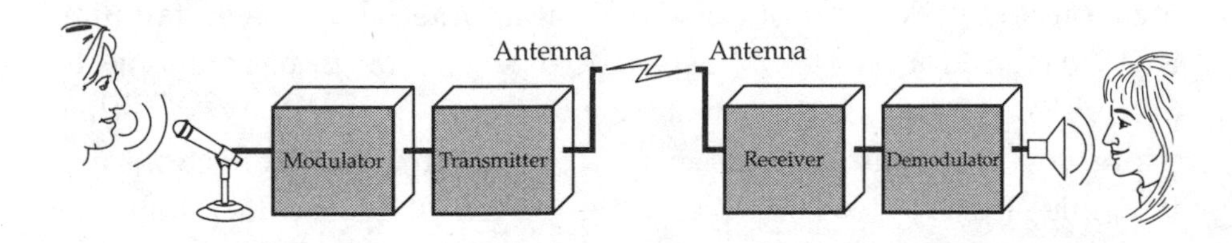

Figure 4-3 Broadcasting Audio Signals

The conversions from electrical energy to electromagnetic energy and back represent another form of transduction where now the transducers are the antennas. Electrical signals propagate through a material by the transfer of energy from electron to electron. For example, copper wires contain large numbers of electrons that are free to move about. In contrast, electromagnetic signals can travel through space, empty or not, as well as through some matter. Electromagnetic waves propagate by themselves, unaided by matter of any kind in contrast to sound waves, which depend upon the mechanism of pressure variations within the materials to propagate.

All wave phenomena, whether they be acoustic or electromagnetic have as their source a vibration or oscillation. If the electrical signals in an antenna were not oscillatory, no radiation of radio waves would take place. The oscillating electrical signals give rise to electromagnetic signals oscillating at exactly the same frequency.

The only thing that distinguishes light from radio waves is its frequency. However, even though light and radio are conceptually the same, for practical reasons it has not always been possible to generate light waves from oscillating electrical signals by a transduction. Until relatively recently, the generation of light always required an intermediate step—as, for example, heating a filament in a light bulb. It was only with the invention of the laser that it became possible to generate light waves with all the flexibility of the other form of electromagnetic waves that we call radio.

Electromagnetic transmission is a phenomenon that is hard to understand intuitively. How can one send signals into the air at one location and pick them out of the air at another? Light is so familiar that we take it for granted. We see it because its reflections from objects are visible. Part of the mystery surrounding radio waves that propagate in exactly the same way is the fact that their reflections are invisible. It may be easiest to resign yourself to the fact that the phenomenon is difficult to understand intuitively, and simply recognize that in some respects they behave like water and sound waves, spreading out as they propagate.

Our understanding of the nature of electromagnetic radiation is based primarily on the work of the great Scottish physicist, James Clerk Maxwell. Maxwell was one of those extraordinary geniuses of the caliber of Newton and Einstein who appear on the scene at rare intervals. While he made significant contributions to many branches of physics in his relatively short lifetime (including the statistical mechanics that I referred to earlier), he is known primarily for his monumental *Treatise on Electricity and Magnetism* published in 1873. Maxwell was the first to recognize that light is electromagnetic in nature—i.e., that it consists of related electrical and magnetic vibrations. He also recognized that all electromagnetic waves, radio and light, propagate at the same speed, the *speed of light*, which is 186,000 miles per second or 300 million (3×10^8) meters per second. Maxwell predicted from purely theoretical arguments that it should be possible to generate electromagnetic radiation, and Heinrich Hertz, the German physicist, was the first to exhibit the phenomenon and thereby demonstrate conclusively the electromagnetic nature of light. Various scientists and engineers (Guglielmo Marconi, the Italian physicist and inventor, is the most famous of these) reduced the idea to practice, and by the beginning of this century, radio or wireless communication had become a reality.

Frequencies and Wavelengths

Another property of waves, whether electromagnetic or acoustic, is their *wavelength*. Wavelength and frequency are related reciprocally: if you know the speed of the wave and its frequency, you can compute the wavelength by dividing the speed by the frequency. Figure 4-4 shows why this is so for sound waves. It shows a sound wave propagating from left to right. A sound wave contains periodic regions of high and low air-molecule densities along the direction of propagation. The distance between successive high or low air-molecule density points of the wave is what we call the length of the wave. The

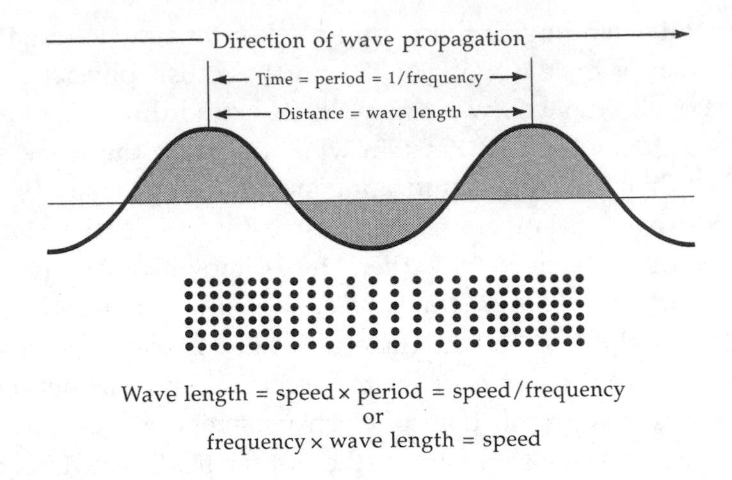

Figure 4-4 Wavelength

wave travels a distance of one wavelength in the time of one period, as shown in the figure. And since electromagnetic waves move with constant speed, the distance the wave travels in this time, the wavelength, is just the period multiplied by the speed. But since the period is the reciprocal of the frequency, the wavelength is the speed divided by the frequency.

Let's see what this means to acoustics. The speed of sound in air is about 1100 feet per second. This means that a 1000-Hz tone has a wavelength of just over one foot, a 10,000-Hz tone has a wavelength one-tenth as large, or slightly under an inch, and a 100-Hz tone has a wavelength of 11 feet. In a home audio system, it is more difficult to achieve good bass response from a small speaker than from a large one. This is because the large speaker is closer in size to the wavelength of the low-frequency tones.

While the formula is the same for electromagnetic waves, the numbers are quite different because the velocities of sound and light are so different. Figure 4-5 shows a chart of a portion of the electromagnetic wave spectrum, labeled by both the frequency and wavelength. Since all electromagnetic waves travel at the same speed,

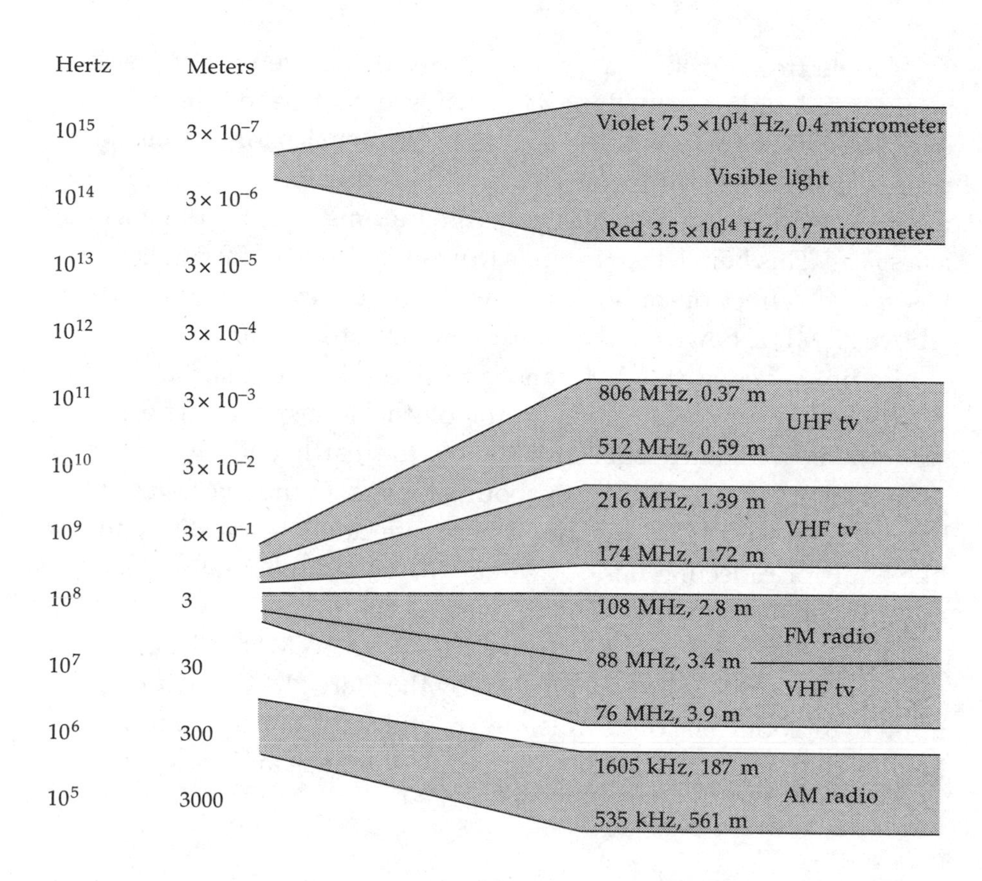

Figure 4-5 A Part of the Electromagnetic Spectrum Extending from the AM Broadcast Band through Visible Light

the higher the frequency, the shorter the wavelength. The familiar AM broadcast band is between 535 and 1605 kHz, with wavelengths in the 300-meter range. The FM broadcast band is between 88 and 108 million Hz (megahertz, abbreviated MHz), with wavelengths in the vicinity of 3 meters. The TV bands are both below and above the FM band. Light frequencies are several orders of magnitude greater, from 350 trillion (3.5×10^{14}) Hz at the red end of the visible spectrum to about 7.5×10^{14} Hz at the violet end, with wavelengths in the vicinity of 0.5 micrometer (millionths of a meter).

All electromagnetic waves, regardless of their length, travel in a straight line unless something is interposed that bends them. However, electromagnetic waves at some frequencies or wavelengths propagate differently than at others. This is because the earth is surrounded by a belt of electrically charged atoms (ions) known as the *ionosphere*. This belt, which extends from an altitude of 50 to 200 miles, has a large effect on radio waves with frequencies below 30 MHz. Above 30 MHz, however, the ionosphere has little or no effect.

Radio waves in the AM band effectively hug the surface of the earth as they propagate as a result of the ionosphere. The signal strength is weakened substantially by the earth and the effective propagation range is limited to about 30 miles. In the *shortwave* band, between 3 and 30 MHz, the ionosphere reflects the waves very much like a mirror reflecting light. Amateur (ham) radio operators use this reflection to transmit their signals over very long distances.

Because the FM and TV bands are above 30 MHz, these waves propagate in straight lines unaffected by the ionosphere. Their range is limited to about 30 to 40 miles by geometrical considerations (the heights of the transmitting and receiving antennas and the curvature of the Earth), commonly called *line-of-sight*.

Amplitude Modulation

The AM frequency band, from 535 to 1600 kHz, is divided into 107 channels, each with a bandwidth of 10 kHz. Broadcasters are assigned one of these channels by the Federal Communications Commission (FCC) for their exclusive use in a particular location. Broadcasters must confine their signals to their own 10-kHz channel to keep from interfering with other broadcasters in the same location assigned to nearby channels. Since the range of an AM station is limited, the same channel can be reassigned to another station in a location that is far enough away to avoid interference between the two stations.

Once an AM broadcasting station is assigned a channel, it must shift its audio signals from their normal frequency range to fit within the bandwidth of the assigned channel. As we noted earlier, this process of introducing audio variations into a high frequency radio signal is called *modulation*. (This should not be confused with the term *modulation* that musicians use to describe changes in tonality. For example, a composer may write a theme in the key of C and then *modulate* the theme or a variant to the key of G. The one thing that the two uses of the term have in common is a frequency change. But the purposes of the change and the nature of the change are quite different.)

The term *amplitude modulation* is very descriptive of the process by which this is done. The broadcast station generates a sine wave at the center of its assigned channel—say at 1000 kHz—called the *carrier frequency*. This term is used because this frequency carries or supports the information contained in the modulated audio signal. Amplitude modulation impresses the audio signal to be broadcast upon the carrier frequency and makes the amplitude of the carrier frequency proportional to the amplitude of the audio signal. Figure 4-6 shows amplitude modulation when the audio signal is itself a pure tone or sine wave. Note that both the top and the bottom of the wave carry the audio information.

Amplitude modulation is generated by a simple multiplication. The modulator is, in essence, an electrical circuit that multiplies the carrier sine wave by the audio signal to be broadcast. The result of this multiplication is the amplitude- modulated signal, as shown in Figure 4-6. What frequencies are in this modulated signal? It turns out that multiplying two sine waves is precisely the same as adding two other sine waves at the sum and difference frequencies of the original sinusoids. Therefore, multiplying an audio signal at 440 Hz by a carrier frequency at 1000 kHz is exactly the same as adding two sine waves at frequencies of 1,000,440 Hz and 999,560 Hz. The top and bottom envelopes of the modulated wave shown in Figure 4-6 are the result of these sum and difference frequencies. Thus, the effect of

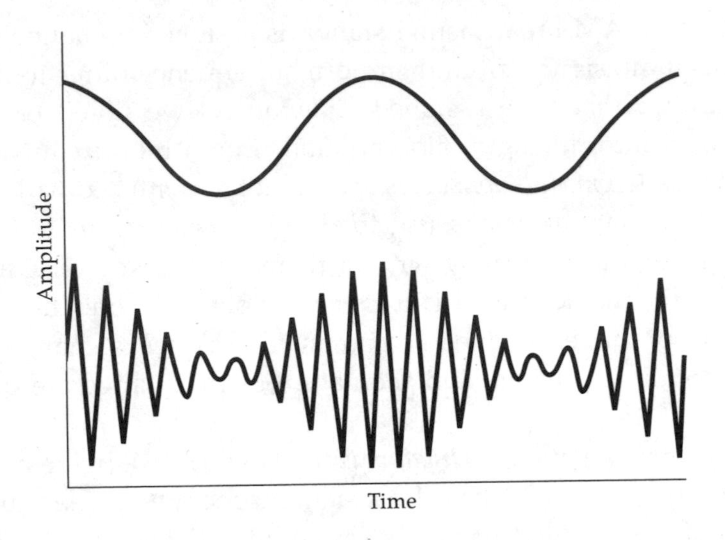

Figure 4-6 Amplitude Modulation

amplitude modulation is to add frequencies both above and below the carrier frequency at a distance equal to the modulating audio frequency. As a result, we can transmit at most a 5-kHz audio band in the 10-kHz AM channel. (It actually has to be a bit less than that to eliminate any chance of interference with adjacent channels.) This explains why AM broadcasting is low-fidelity. The demand for channels is so great and the width of the band so limited that only 10 kHz can be allocated to a channel.

Frequency Modulation

Frequency modulation was developed as a way to improve the quality of radio broadcasting. AM broadcasting has the drawbacks of its limited audio bandwidth and susceptibility to noise. (For instance, an electrical storm can be devastating to AM reception.) FM achieves its improved performance by using more bandwidth per channel. The band dedicated to FM is 20 MHz wide, extending from 86 to 106 MHz.

The channels in this band are 200 kHz wide, 20 times wider than the 10-kHz channels in the AM band. If we were to use AM in this frequency band, we would only need 40-kHz channels to permit the broadcasting of full-frequency audio. FM channels are so much wider because the noise resistance of FM depends upon the amount of bandwidth used—the more the better.

Figure 4-7 shows a comparison of the two modulation processes. The FM waveform looks quite different from the AM waveform. In frequency modulation, the frequency rather than the amplitude of a carrier sine wave is modified in proportion to the amplitude of the

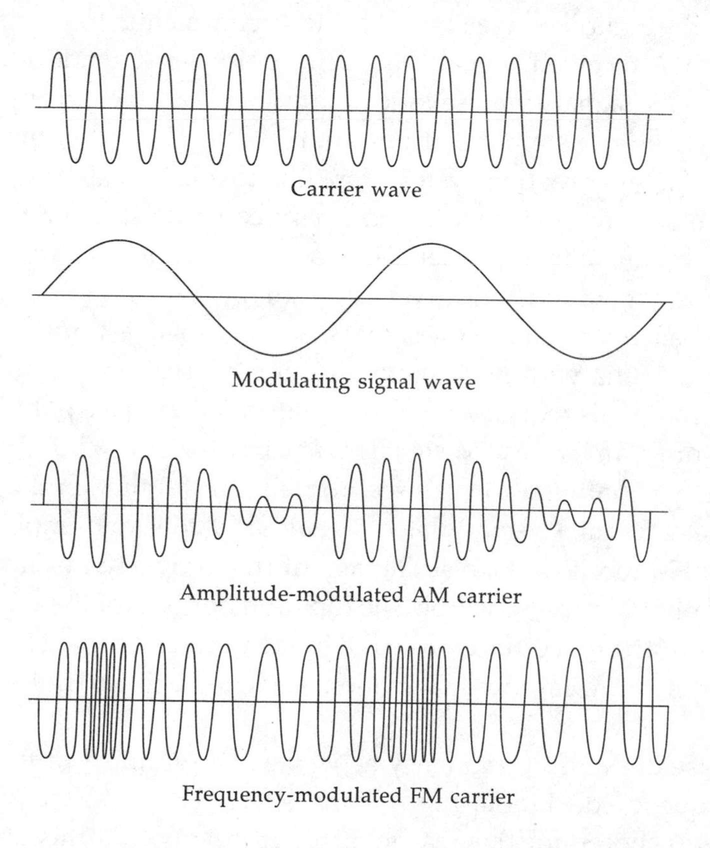

Figure 4-7 Comparison of Amplitude Modulation and Frequency Modulation

audio signal. At the peak of the modulating audio signal, the carrier frequency is increased by the largest amount; and at the trough of the modulating signal, the carrier frequency is decreased by the largest amount. Thus, the size of the maximum frequency deviation is an important contributor to the amount of bandwidth necessary to transmit the signal. For example, in the standard broadcast FM system, the carrier frequency is deviated on either side by 75 kHz to support the maximum amplitude of the audio. In a narrower bandwidth system, a carrier deviation of 10 kHz might be used to support the same audio amplitude.

The bandwidth determines the effect of noise on the system. The farther the radio receiver is from the broadcasting transmitter, the weaker the received signal. Since the noise level tends to remain constant, the signal-to-noise ratio at your receiver depends upon how far away the transmitter is from your home. Therefore, the acceptability of a broadcasting system depends to a large extent upon how well the system performs in the presence of noise over the entire geographic area that it is intended to serve, including the fringe areas.

Noise generally is additive. The signal entering the receiver is a reduced-energy replica of the signal that was broadcast from the transmitting antenna with noise, sitting on top of it. Therefore noise perturbs AM signals that carry the audio information in amplitude variations more than it does FM signals that carry the audio information in frequency variations. You can see from Figure 4-7 that, in the absence of noise, a frequency-modulated signal has a constant amplitude envelope. FM receivers take advantage of this property by clipping off any amplitude variations from the tops and bottoms of the waveforms before extracting the audio from the frequency variations. They can do this because any amplitude variations have to be due to noise of some kind.

Noise does affect frequency as well as amplitude, but the greater the frequency deviation, the less the effect. The 200-kHz frequency band was chosen for this reason. Broadcasting within this range permits the transmission of audio with frequency deviations large enough to provide significantly more immunity from noise than does AM transmission. However, the greater the dynamic range (range of amplitudes) in the audio, the greater the required frequency deviations. Because of this, even with 200-kHz-wide channels, the frequency range of the audio must be limited to 15 kHz and the amplitude range of audio signals must be compressed before modulation. For this reason, the fidelity achievable in FM broadcasting, while much better than that of AM, is still below that achievable with high-quality analog recording and certainly below that typically achieved with digital recording.

Many FM stations transmit stereophonic signals. This simply means that they broadcast two audio signals instead of one, each obtained from a different set of microphones placed to the left and right of the signal source. Interestingly, they are able to do this in approximately the same bandwidth. A stereophonic receiver can distinguish between these two signals and direct them to separate audio channels that drive separate loudspeakers. The scheme used for the stereo broadcasts is also compatible with monophonic receivers, because it allows them to receive both signals as if they originated in the same microphone and directs them to a single speaker. The stereo effect, of course, is not present.

Television

Television broadcasting is a form of radio broadcasting. Actually, we should call radio broadcasting, *audio broadcasting*, and television broadcasting, *audio/video broadcasting*. Sending television means sending signals that represent both pictures and sound. A video camera sweeps across its field of view and generates an electrical signal that represents the light intensity as observed by the camera. This electrical signal is a waveform that is proportional to the brightness of the image that the camera sees as it scans across the scene. At the same time, a microphone converts the sound into another waveform. From there, the process is much the same as with radio broadcasting. The video and audio waveforms modulate their own individual *subcarriers* and these are combined on a final carrier frequency in one of the channels in the TV band.

The essential difference between the video and audio signals is their bandwidth. High-quality FM audio broadcasting needs 200 kHz of bandwidth; 6 million hertz (Megahertz or MHz) is required to send a color TV broadcast. Since the audio in a television broadcast occupies even less bandwidth than does the FM broadcast, virtually all the bandwidth is consumed by the video signals. One of the most important factors that determine this bandwidth is the definition or resolution of the picture. In the U.S. standard, the camera sweeps across the picture 525 times as it moves from the top to the bottom of the scene. Although that resolution is really quite crude, it is adequate for viewing on the usual screen found in most homes. But large-screen projection systems show up the crudity of this resolution, certainly when it is compared to normal optical motion pictures. For example if the projection screen is four feet on a side, there are only about 10 lines per inch. Nevertheless, most of us have become accustomed to this level of resolution. A higher resolution process called *high-definition TV* (abbreviated *HDTV*), with about twice the number of lines is being developed. We'll discuss this later on when we discuss digitized video.

You have undoubtedly noted that the audio quality of your TV is quite poor when compared to FM reception through a high- fidelity system. There are two reasons for this. First, the television audio is transmitted in a narrower bandwidth and is usually monaural. Second, the TV receiver manufacturers have used very low-quality audio systems to keep the cost of the receivers as low as possible. The resulting quality is hardly better than that of a table-top radio tuned in to an FM broadcast. More recently, some TV stations have begun to broadcast their audio in stereo and some TV receivers have been marketed with either upgraded audio or a convenient outlet terminal that you can hook up to your audio system.

Of course, to be philosophical about it, low-quality audio is well matched to the vast bulk of today's TV programming. After all, what benefit might be attached to the use of hi-fi audio with the average sitcom? But every once in a while, there is a televised musical event of such quality as to merit a *simulcast* in which a network (usually Public Television) will broadcast the audio portion of the program in stereo over an affiliated FM station. With such a simulcast, you simply turn down the volume on your TV and listen to your high-quality radio. These simulcasts will become less important as high-quality stereo TV audio broadcasting becomes more widespread.

Frequency and Bandwidth

The bandwidth of a radio channel is highly dependent upon the frequency band in which the channel is located. For example, a single TV channel has a bandwidth six times wider than that of the entire AM band. Similarly, the entire shortwave band—which can support long-distance transmission by ionospheric reflection—is only 27 MHz wide, less than that of 5 TV channels. It was inevitable, then, that the search for more bandwidth led to the development of the technology that permitted the use of the higher frequency bands, first the microwave regions (the frequencies above 1000 MHz, or 1 gigahertz—-abbreviated GHz) of the spectrum, and later the optical regions.

The development of radar during World War II provided the microwave technology that began to be applied to telecommunications soon after the war. AT&T covered virtually the entire country with microwave relay circuits to replace its lower capacity cables. (A microwave relay is a chain of individual line-of-sight microwave radio links spanning a long distance.) Later, when the communications satellite was developed, it was used to span even longer distances for both the telephone networks and television. The satellite made a spectacular difference for transoceanic transmission where microwave relay is not a factor. Until the first Intelsat in the 1960s, all overseas telephone traffic was carried by the relatively low-capacity undersea cables, and there was no way of supporting the large bandwidths required for television. The television networks were forced to carry tapes of fast-breaking news events by jet aircraft. The communications satellite using the microwave bands has had an enormous impact on the ability of the broadcast media to bring virtually instantaneous coverage of world events into our homes.

The optical band with its enormous bandwidth was next to be exploited. Since the atmosphere presents such a severe limitation to optical propagation, it was clear from the start that optical transmission had to be by cable. It has already had an enormous impact on the telecommunications networks, first for domestic and, later for overseas routes.

Analog Recording

Recording is a special kind of communication—communication with delay, as opposed to live or *real-time communication*. The entire recording and playback process constitutes the communications channel. Like any other communications channel, the recording channel introduces distortions. It may introduce frequency limitations and it may add noise. In a vinyl record, a replica of the actual audio waveform is cut into the plastic. In an analog tape recording, the replica of the audio waveform is in the form of magnetization proportional to the waveform amplitude. In both cases, the playback mechanism converts the frozen image on record or tape back into an audio signal that is ultimately reconverted to an acoustic signal through a loudspeaker.

We can think of the entire chain of equipment used in the recording/playback process as having a frequency response, analogous to the way that a telephone line has a frequency response. This frequency response determines how much of the original waveform is passed on to the listener. Similarly, the recording channel introduces noise, most of it in the playback mechanism. One example of such noise is record scratch. A pickup needle makes physical contact with the record surface and causes a slight amount of abrasion. The effect is cumulative, and after the record is played enough, the surface noise caused by the successive abrasions interferes with the pleasure of listening. That means it's time to buy a new record to create a new channel.

This connection between communicating and recording, while only of academic interest in the analog case., has become of practical significance in the development of digital recording systems. We shall see this later on.

5

The Wonderful Symmetry of Frequency and Time

With a few strokes of his pen a skillful cartoonist can capture an idea that a writer can labor over in paragraph after paragraph. Thus, in Figure 5-1 Garry Trudeau expresses with eloquent humor one of the major differences between digital and analog communications: the fact that when information is expressed in digital form, the digits, although representative of the information source, have lost the direct connectivity to the source that is always present in analog communications.

Indeed one of the things that makes analog communications so appealing is precisely the naturalness of this connectivity. Analog information transmission is clear and intuitive: the acoustic signals produced by the sources of speech or music consist of combinations of frequencies derived from the natural vibrations of the acoustic source;

Figure 5-1 A Catchy Digital Tune (DOONESBURY Copyright 1988 G. B. Trudeau. Reprinted with permission of Universal Press Syndicate. All rights reserved.)

the electrical waveforms are proportional to these acoustic signals; finally, the transmitted signals are related directly to the electrical waveforms and, of special significance, their bandwidths have a clear relationship to the information carried by the audio signals.

Communicating Digits

Digital transmission is similar to analog transmission in many ways. To communicate data in digital form over communications media such as telephone lines or radio circuits, the digits have to be represented by waveforms that are compatible with the transmission media. A stream of binary digits is converted to a stream of waveforms, one waveform representing a 1 and another representing a 0. These waveforms, like all waveforms in this world, have frequencies and bandwidths and are transmitted through the channel in ways that are similar to the transmission of analog waveforms.

One important difference between digital and analog transmission is that the signals carrying digital information are simply numbers, 0s and 1s, mathematical abstractions. The digits may represent computer data files, facsimile, music, speech, television, or anything else. But once the information source has been converted into digital form, a stream of digits is a stream of digits regardless of its origin or significance. This very sameness in digital transmission blurs the connection between the transmitted signals and the source information that is always present in analog communications. Fortunately these relationships that seem anything but natural at first glance become less tenuous once the fundamental properties of waveforms are explored.

Like an analog signal, a digital signal must also be tailored to the channel that will transport it. To avoid losing information, the frequencies contained in the signal waveforms must fall within the bandwidth of the channel. Thus, to send digits over a telephone line, the waveforms representing the digits must have bandwidths within the telephone line bandwidth.

An example of signals that might be used on a telephone line is shown in Figure 5-2. It is the same scheme that I used in Chapter 2 to introduce digital communications. In this scheme, a 1 is represented by a signal pulse T seconds long and a 0 is represented by the absence of a pulse for the same T seconds. One binary digit is sent every T seconds, making the data rate $1/T$ bits per second. If the pulse is 1

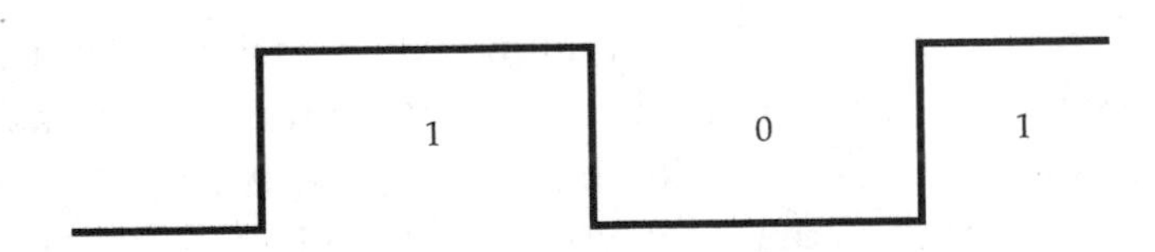

***Figure* 5-2** On-Off Signaling

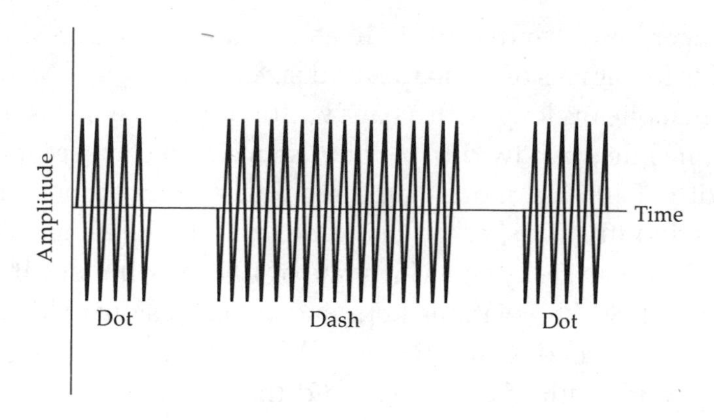

Figure 5-3 Radio Telegraphy

millisecond (one-thousandth of a second, abbreviated ms) long, then the data rate is 1000 bps. The shorter the pulse interval *T*, the higher the data rate.

Sending Morse code over a radio circuit is similar. Figure 5-3 shows examples of the signals that are typically used. A long pulse of a radio-frequency carrier represents a dash, and a short pulse of the same carrier represents a dot. (We could substitute 1 and 0 for dash and dot, respectively.) In this case, too, the shorter the pulses, the higher the data rate.

But changing the signal width to vary the data rate also changes the bandwidth. The nature of this relationship between pulse duration and bandwidth is at the heart of digital signaling.

Bandwidth and Signal Duration

The general rule that relates signal (pulse) duration and bandwidth is reciprocality: the bandwidth is the reciprocal of the signal duration. If you know the duration of a pulse, then you also know its bandwidth. A signal one second long has a bandwidth of one hertz, and a signal 10 times longer has a bandwidth 1/10 as large. This reciprocal relationship holds for signals of all shapes and sizes.

A signal one ms long with a bandwidth of 1 kHz will easily fit into a telephone line's bandwidth. In contrast, a signal 0.1 msec long has a bandwidth of 10 kHz, more than three times wider than a telephone line's bandwidth. Thus, when using the on/off signaling scheme of Figure 5-2, we can easily signal at a rate of 1000 bps, but not at a rate of 10,000 bps. (The abbreviations kbps and Mbps designate kilobits and megabits per second, respectively.) When these on/off signals are used, the bandwidth of the signals and the data rate are numerically the same. This has led to the use of the terms data rate and bandwidth interchangeably when speaking of digital channels. This loose terminology is neither accurate nor desirable, because, there are other signaling schemes and noise conditions in which the data rate, while always related to the bandwidth, does not equal it. Therefore, I will preserve the distinction between the two terms.

Reciprocal relationships like this one between time and frequency are found elsewhere. For example, the view angle of a lens is related to the diameter of the lens reciprocally. A large lens has a small view angle and vice versa. A large telescope such as the instrument on Mount Palomar obtains its power to see very distant objects at the cost of a very narrow field of view, just as communications signals obtain their ability to represent high-rate information at the cost of higher bandwidths. The same thing is true of dish antennas commonly used for communicating with satellites: the larger the antenna diameter, the greater its transmitting or receiving capability, but the narrower the beam formed by the antenna.

Quantum mechanics describes the motion of atomic particles. Heisenberg's famous uncertainty principle states that it is impossible to measure the position and velocity of a particle *simultaneously* with perfect accuracy. The greater the accuracy by which the position is measured, the poorer the accuracy by which the velocity can be determined, and vice versa. The two accuracies or uncertainties are related reciprocally.

All these examples of the reciprocality relationship obey the same kinds of mathematical formulas, even though the physical principles seem markedly different. The mathematics has its origin in the work of the nineteenth century French mathematician, Jean-Baptiste Joseph Fourier. This mathematics, called Fourier Analysis, describes the behavior of periodic signals. I introduced it earlier in discussing the harmonic relationships produced in musical vibrations.

Fourier's mathematics was motivated by his interest in explaining heat conduction phenomena. In contrast, most pure mathematics in recent years has been an intellectual pursuit inspired by its inherent interest rather than motivated by the need to solve physical problems. Indeed, David Hilbert, one of the giants of early twentieth century mathematics, when asked to address a Joint Congress of Pure and Applied Mathematics to break down the supposed hostility between the two groups, began his talk in the following way:

> We are often told that pure and applied mathematics are hostile to each other. This is not true. Pure and applied mathematics are not hostile to each other. Pure and applied mathematics have never been hostile to each other. Pure and applied mathematics will never be hostile to each other. Pure and applied mathematics cannot be hostile to each other, because, in fact, there is absolutely nothing in common between them.

Not so in previous generations. Before the modern age of specialization, mathematicians and physicists were often indistinguishable. Fourier contributed to both fields. It is also of interest that he was not only a brilliant and eclectic scientist but also a man active in the affairs of the world. He lived in a time of great political and military turmoil in his native country, and he participated in these events as they unfolded both in the early days of the French revolution (to his credit, he is said to have opposed its bloodier aspects) and later during the reign of Napoleon.

According to Fourier's theory, a periodic waveform is a periodic waveform regardless of its origin. Periodic waveforms that arise from nonmusical sources have the same characteristics as periodic musical waveforms. Because the underlying mathematics is the same, we can take principles that apply to music and speech and apply them elsewhere beyond the acoustical realm, using the insights derived from the musical connection. They provide the link connecting digital to analog communications. But even aside from this, Fourier's analysis has an esthetic appeal that transcends any application to specific practical problems.

Fourier Analysis of Pulsed Waveforms

Pulsing a Tuning Fork: Equal On/Off Times

We will get to the heart of the reciprocality relationship through Fourier's theory using pictures in place of the mathematics.

To do this we will use musical examples involving the tuning fork. The tuning fork is hardly the most interesting instrument musically. However, it has the unique property stated earlier of producing a sound that is very close to a pure tone. Its waveform is, therefore, very close to a pure sine wave. It is this property, in fact, that makes it so useful to a piano tuner.

Let's imagine an idealized experiment in which we sound a 100 Hz tuning fork for precisely 1 second, stop it for 1 second, restart it for one second, and so forth. The result is a rhythmic sequence of tone pulses that, when played on a violin or other stringed instrument, is called *tremolo*. Its waveform, shown in Figure 5-4, looks very much like the Morse code waveform of Figure 5-3. What can we say about its constituent frequencies? We know that there must be frequency components in addition to 100 Hz, because it sounds different from the sound produced when the tuning fork is allowed to vibrate indefinitely. While the tuning fork is being sounded, its frequency is 100 Hz. But it is only being sounded half the time.

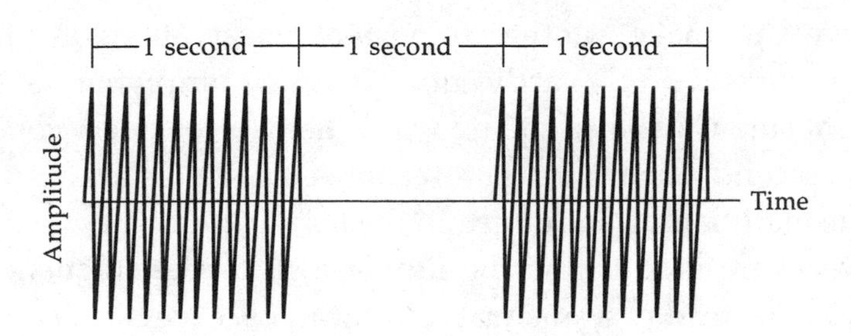

Figure 5-4 Pulsed Tuning-Fork Waveform

This is where Fourier's theory comes in. Recall that every musical instrument has a vibration mechanism, e.g., the string of a violin or piano, or a column of air in a brass or woodwind instrument, that produces a set of higher harmonics along with the fundamental frequency or pitch. When a note is played on an instrument, say an oboe, the resulting audio signal is the sum of the tones at all the harmonic frequencies of the fundamental of the note being played. It is possible, in principle, to compute the amplitudes and phases of these harmonics simply by knowing how the column of air within the oboe vibrates. But what is possible in principle is not always practical. The much less laborious technique developed by Fourier allows us to compute these amplitudes and phases by analyzing the musical waveform itself; i.e., by decomposing the waveform into its constituent harmonics. Fourier's theory can also be applied to determine the harmonic content of periodic signals when the signals are not derivable from the vibrations of a musical instrument. The formula for computing the amplitudes and phases of the harmonics is just the same in this case.

The waveform shown in Figure 5-4 is periodic. The fundamental frequency is 0.5 Hz, since the pulses occur at the rate of one every two seconds. The higher harmonics are 1, 1.5, 2, 2.5 Hz, etc. These harmonics must continue at least up to 100 Hz, the tuning fork frequency, and even beyond, because the rapid beginnings and endings of the pulses must indicate the presence of some much higher frequencies.

But we can make another important observation: turning the tuning fork on and off is exactly equivalent to multiplying the 100-Hz tuning-fork sine wave by a square wave that is 1 for one second, then 0 for one second, then 1 for one second etc. This is an example of amplitude modulation, as defined in the last chapter. The difference is that now we are modulating the tuning-fork carrier frequency by a square wave instead of a sine wave. You can observe the similarity in Figure 5-5, which shows the result of modulating or multiplying the tuning-fork sine wave by a sine wave with a frequency of 0.5 Hz, the fundamental frequency of the square wave. While you could not do this with an actual tuning fork, you could do it with an electronic music synthesizer, which can create any combination of tones. The modulated tone would sound like a 100-Hz tone with a pulsating wobble.

When a carrier frequency's amplitude is modulated by a sine wave, the resulting waveform contains three frequencies: the carrier frequency and the frequencies obtained by adding and subtracting the modulating frequency from the carrier. Since in this case the carrier is the 100-Hz tuning fork tone and the modulating frequency is the 0.5-Hz sine wave, the waveform contains the frequencies 99.5 Hz and 100.5 Hz in addition to the 100-Hz carrier frequency.

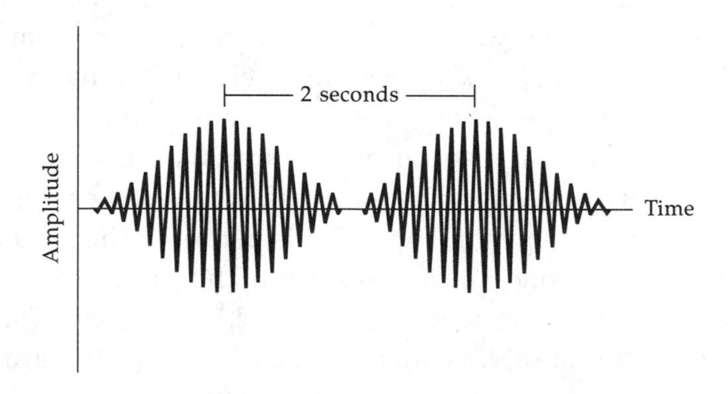

Figure 5-5 Tuning-Fork Waveform Modulated by a Sine Wave

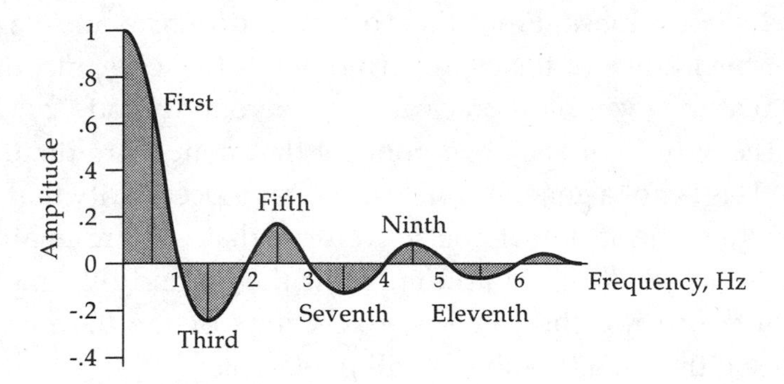

Figure 5-6 The Frequency of Components of a Square Wave

More generally, when you amplitude-modulate a carrier with an audio signal containing any number of different frequencies, you obtain the sum and difference frequencies of the carrier and all the audio components. For this reason, an audio band containing all the frequencies up to 5 kHz, when amplitude- modulating a carrier, produces all the frequencies 5 kHz above and below this carrier, thereby filling the entire allocated band of 10 kHz. This gives us a clue to the nature of the frequency components of the pulsed tuning fork waveform. According to Fourier, the modulating square wave will contain all the harmonics of its 0.5 Hz fundamental. The pulsed tuning-fork waveform should contain, in addition to the 100 Hz carrier, all the sum and difference frequencies of the carrier and each of these harmonics.

The amplitudes and phases of these various harmonics are calculated from Fourier's formulas. The results can be expressed in the useful pictorial form of Figure 5-6. The vertical axis is amplitude, and the horizontal axis is frequency. Note the shape of the curve. It has a large positive half lobe beginning at the left end with subsidiary full lobes alternating in sign and decreasing in amplitude as the frequency increases. The amplitude of any particular harmonic of the square wave is given by the height of the curve at the frequency of that harmonic. The fundamental frequency of 0.5 Hz falls in the middle of the first lobe, and the second harmonic occurs at the junction of the first and second lobes. From then on, the odd harmonics fall at the peaks of the lobes and the even harmonics at the lobe junctions. This tells us that the even harmonics all have zero amplitude, or, in other words, the only frequency components that matter are the odd harmonics. This is not a general conclusion. It is a peculiarity that follows directly from the fact that the waveform that we are analyzing is square. As we shall see in a moment, if the on and off times of the tuning fork are not the same, the positions of the harmonics will change, and the even harmonics will play a role.

The fact that the amplitudes of the lobes decrease fairly rapidly as the frequency increases means that the low-numbered harmonics, with their larger amplitudes will contribute most to the square wave, and we can probably neglect most of the higher harmonics.

Now let's return to our on/off tuning-fork waveform. The preceding discussion tells us that the square-wave modulating signal can be represented by a few of the lower harmonics of 0.5 Hz. It follows that modulating the tuning-fork sine wave with the square wave is just the same as modulating it by the sum of these few harmonics. Therefore, we would therefore expect the frequency components of the tuning fork modulated by the square wave to be the sum and difference frequencies of the carrier and these harmonics. Thus, the frequency components of the pulsed tuning fork are:

Carrier	100 Hz
First harmonic	99.5 and 100.5 Hz
Second harmonic	99.0 and 101.0 Hz
Third harmonic	98.5 and 101.5 Hz
Fourth harmonic	98.0 and 102.0 Hz
Fifth Harmonic	97.5 and 102.5 Hz
(etc.)	

I have used the word *harmonic* in the table above somewhat loosely. The first harmonic is the sum and difference of the carrier and the fundamental (first harmonic) of the square wave. And the table in-

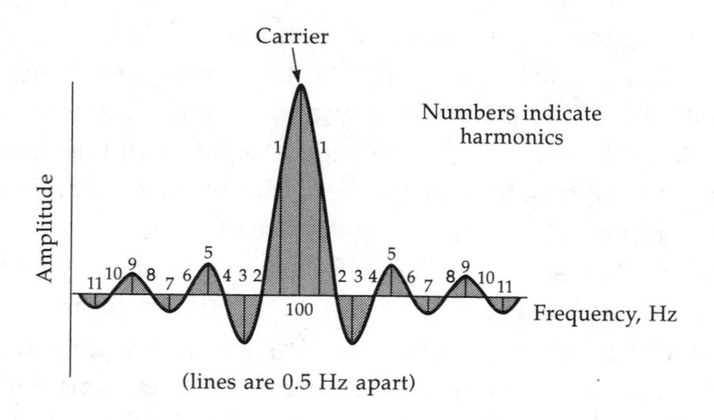

Figure 5-7 Amplitudes of the Harmonics of a Pulsed Tuning Fork

cludes the even harmonics even though we know that their amplitudes are zero. The reason for this will become clear a little later.

The amplitudes of these modulated harmonics are shown in Figure 5-7. The curve looks similar to the curve in Figure 5-6 because it is obtained by laying the curve of Figure 5-6 on either side of the 100-Hz carrier, the pictorial equivalent of taking the sum and difference frequencies. Now the initial half lobe of Figure 5-6 becomes a central lobe in Figure 5-7, and the subsidiary side lobes appear symmetrically on either side of the carrier. Since the Figure 5-7 picture is another view of Figure 5-6, the harmonics fall at the same places on the curve *relative* to the carrier frequency. The first harmonic frequencies fall within the central lobe. All the other odd harmonics fall on the peaks of the successive side lobes, and the even harmonics, as before, fall on the lobe junctions. These junctions are commonly called *zero-crossings* for obvious reasons.

The position of these zero-crossings is very significant. The first occur 1 Hz (the reciprocal of the pulse width) on either side of the carrier tone. This means that the central lobe has a width from zero-crossing to zero-crossing of 2 Hz or twice the reciprocal of the pulse width. The other lobes have widths one-half that of the central lobe.

If the tuning fork were allowed to sound indefinitely, its waveform would contain a single frequency of 100 Hz, a single line at the peak of the central lobe in the picture of Figure 5- 7. Thus, the effect of the pulsing is to replace the single line with a sequence of lines clustered around the tuning-fork frequency. The width of the central lobe is, in some sense, representative of the bandwidth of the pulse.

The waveforms shown in Figure 5-8 are obtained by synthesizing the pulsed tuning-fork waveform from its constituent harmonics with the amplitudes as given in Figure 5-7. Trace *a* shows the sum of the carrier and the first harmonics; in trace *b*, the third harmonics are added; and in trace *c*, all the harmonics up through the fifth. Note that with as few as three frequency components, the general shape of the square wave becomes evident. As more harmonics are added, a better

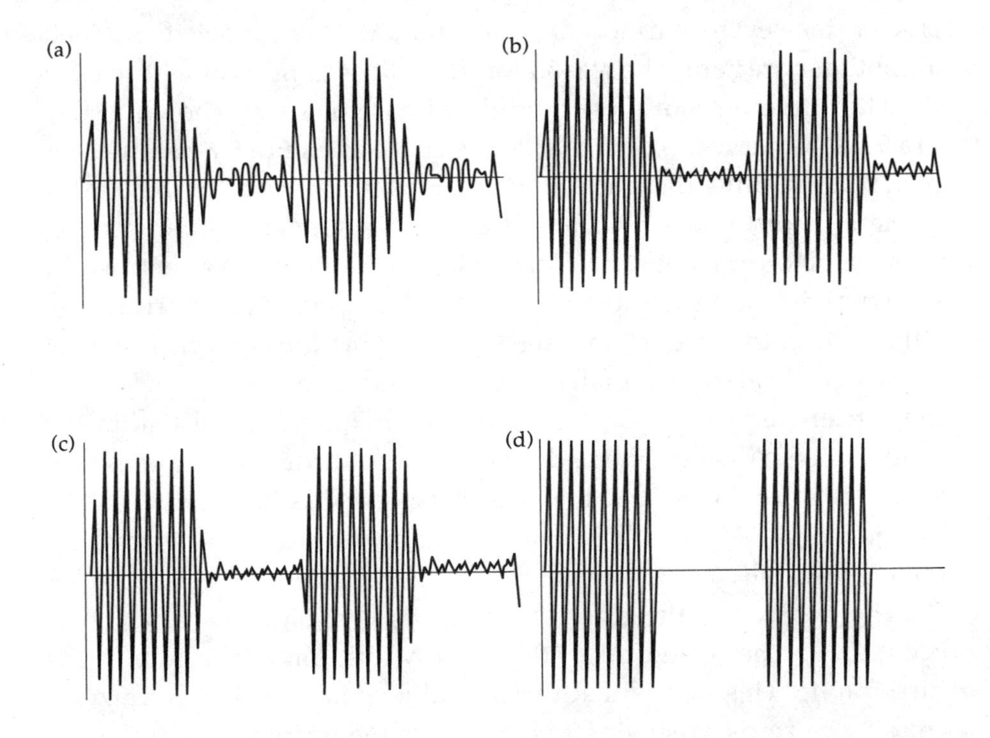

Figure 5-8 Reproduction of the Pulsed Tuning-Fork Waveform

approximation of the square wave is obtained. If we were to use all the harmonics, we would reproduce the original waveform shown in the trace *d*.*

Pulsing a Tuning Fork: Unequal On/Off Times

To gain more insight into the problem, we will modify our hypothetical tuning-fork experiment slightly. Instead of sounding the tuning fork for 1 second at 1-second intervals, we will sound it for 1 second and leave it off for 3 seconds, producing the waveform shown in Figure 5-9. It is similar to the original waveform shown in Figure 5-4, except that it is rectangular instead of square with a period of 4 seconds instead of 2. With these modified parameters, its fundamental frequency is now 0.25 Hz instead of 0.5 Hz. Using the same rule as above, the frequency components are:

Carrier	100 Hz
First harmonic	99.75 and 100.25 Hz
Second harmonic	99.50 and 100.50 Hz
Third harmonic	99.25 and 100.75 Hz
(etc.)	

The amplitudes associated with each frequency component are also shown in Figure 5-9. The envelope of the curve is identical to that shown in Figure 5-7. But now the spacing of the frequency components is half that in the first example, because the period of the wave has doubled. Now the first three pairs of harmonics occur within the central lobe, as compared to the first pair in the previous example.

* Figure 5-7 shows how this particular waveform may be reproduced from its Fourier components using amplitude information only. When dealing with more complex waveforms than this one, phase information is usually required as well.

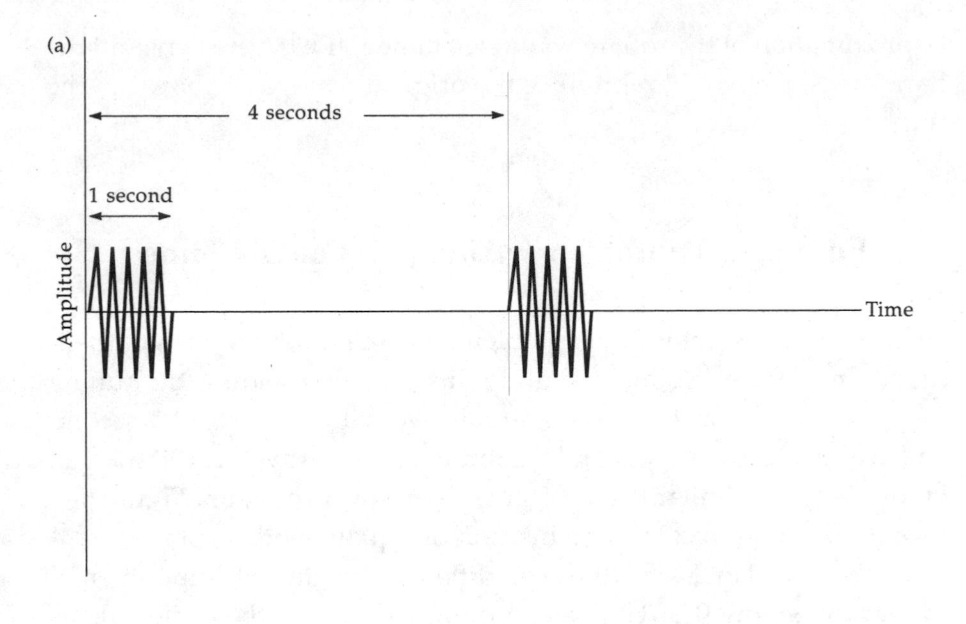

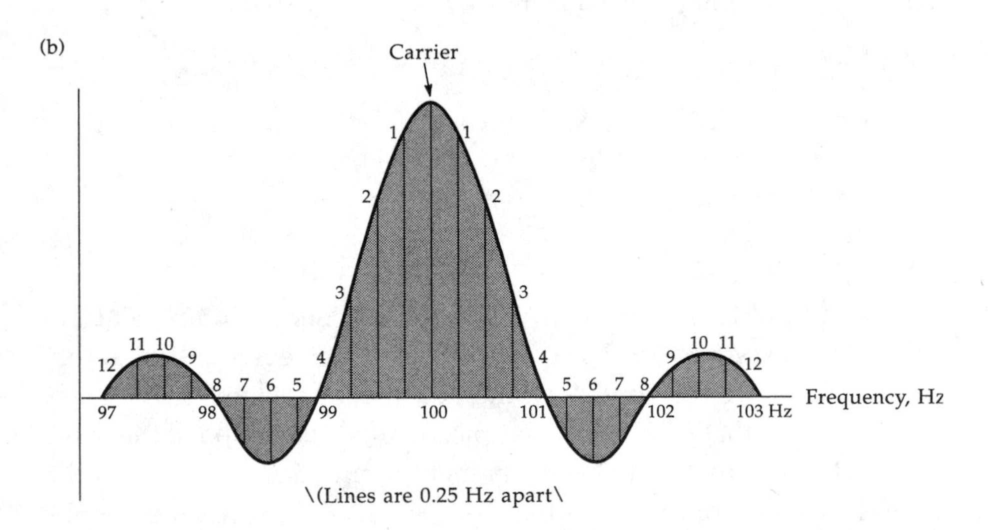

Figure 5-9 Pulsed Tuning-Fork Waveform with 4-Second Period

And the fourth, eighth, twelfth, etc. harmonics, rather than all the even harmonics, occur at the zero crossings. If we attempted to synthesize the waveform by adding up its constituent harmonics, as we did before, we would have to use twice as many to obtain the same degree of approximation.

If we repeated the experiment, each time separating the pulses of tuning-fork vibration by greater and greater intervals, the harmonics of the waveforms would move closer and closer together—but always with the same envelope shown in Figure 5-7. In the extreme case where the pulses are very far apart, the frequency components are so close to one another that they form a continuum completely covering the envelope of Figure 5-7.

In each experiment, the location of the harmonics depends upon how often the tuning fork is sounded—-the more often, the farther apart the successive harmonics. But the curve that determines the amplitude of the harmonics (usually called the *spectral envelope* or simply the *spectrum*), wherever they occur, is the same, because this curve depends only upon the *shape* of the tone pulse and not at all upon how often the pulse occurs. This allows us to separate the effects of pulse width from the effects of pulse periodicity. A rectangular pulse of length T *always* has the same spectral envelope, regardless of the rate at which the pulses occur. The periodicity of the pulse lets us locate the harmonic lines on this spectral envelope.

Interpretation of Pulsed Waveforms

There is a great deal that we can infer from the discussion in the previous sections. First, we have demonstrated the reciprocal relationship discussed earlier—i.e., a pulse T seconds long has a bandwidth of approximately $1/T$ Hz. This follows directly from Figures 5-7 and 5-9, which show that, regardless of how often the pulse occurs, the total width of its spectrum between zero-crossings is $2/T$. I used the qualifier *approximately* because the width can be estimated in different ways. The width between zero-crossings is one way. Another, perhaps more reasonable, way recognizes the fact that the amplitudes of any harmonics occurring near the zero-crossings are small, and approximates the effective width of the spectrum by half the total width of the central lobe.

This means that a very long pulse has a very narrow spectrum, and vice versa. A very long pulse would result from striking a tuning fork and allowing it to vibrate for a long time before stopping it. In this case, the spectrum of the waveform is the frequency of the tuning-fork vibration with a very narrow envelope—in essence, the single tuning-fork frequency that a piano tuner finds useful.

By the same token, if a pulse is very narrow, its spectrum is very wide. A periodic series (or train) of very narrow pulses results in a spectrum wide enough for a very large number of harmonics to fit in the central lobe. Figure 5-10 shows this pictorially. Figure 5-10a shows the pulsed tuning fork waveform with pulses 0.1 ms wide occurring every 10 ms, corresponding to a 100-Hz rate. Figure 5-10b shows the spectrum of this pulse train. The central lobe has a total width (zero-crossing to zero-crossing) of 20 kHz (2/0.0001 s). Therefore, the first 100 harmonic pairs of the waveform fit in this central lobe. We saw waveforms like this one in Chapter 3 during the discussion of speech. Recall that when the vocal cords vibrate, they send out periodic puffs of air at the fundamental frequency of vibration. The spectrum of these pulses is a train of harmonic frequencies of about the same amplitude. We showed an example of this in Figure 3-14. The reason that there are so many harmonics of about the same amplitude is that the puffs of air are short and, accordingly, their spectrum is wide.

Every pulse shape has a particular spectral envelope associated with it, determined by Fourier's formulas. The specific spectral envelope of Figure 5-7 that we have been using for all our examples to this point applies only to rectangular pulses. These contribute the high-frequency components of the beginning and end of the pulse. Such pulses have an idealized shape with perfectly vertical sides, which implies that the tuning fork is turned on and off instantaneous-

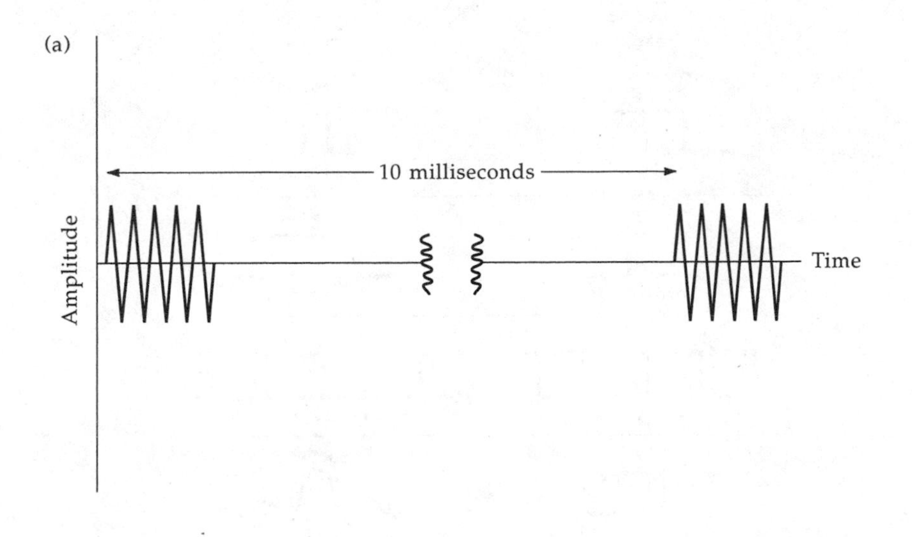

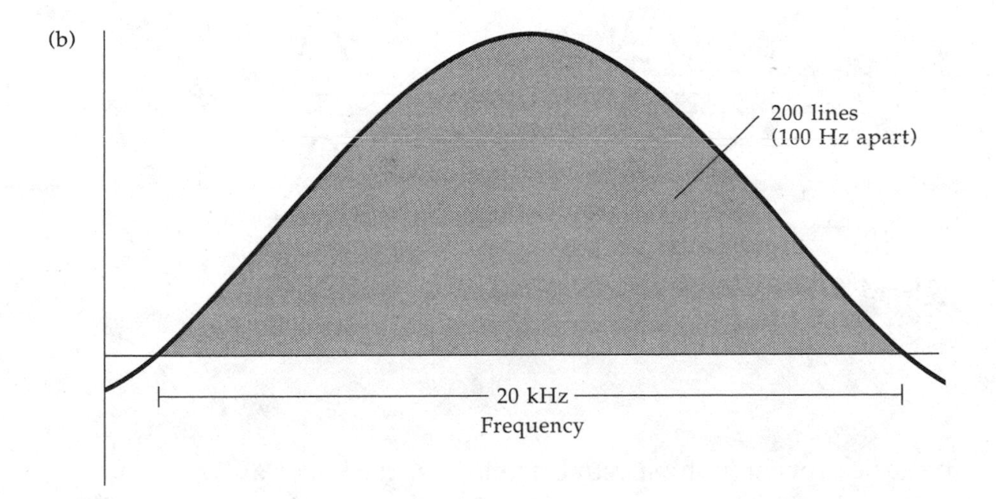

Figure 5-10 Spectrum of Widely-Separated Narrow Pulses

ly. In musical terms, it implies infinitely fast *attacks* and *releases*, the terms commonly used to characterize the beginnings and ends of musical phrases. It is easy to see that the steeper the sides of these waveforms, the greater their high-frequency content. These high fre-

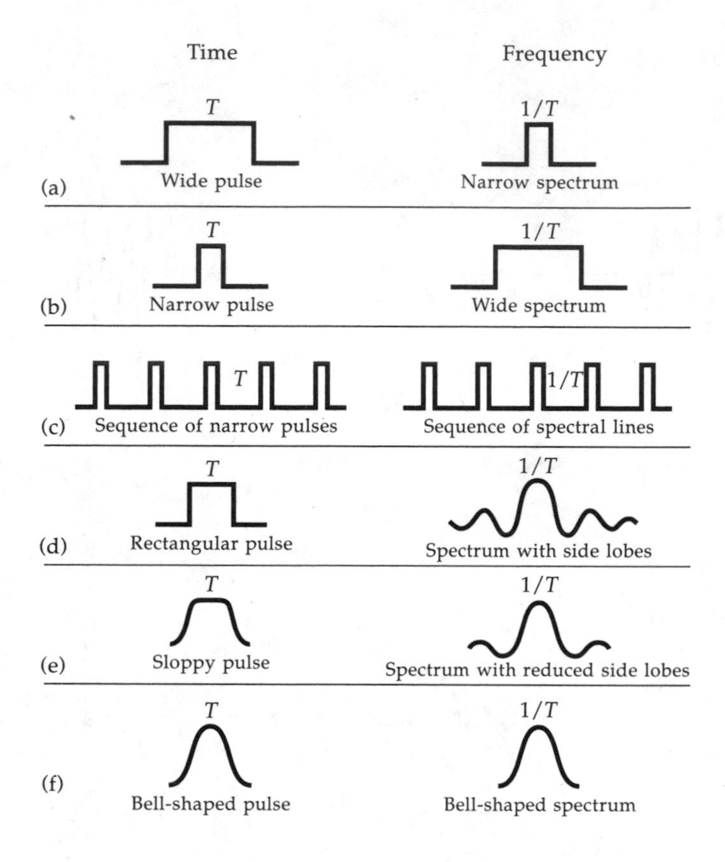

Figure 5-11 Time and Frequency Reciprocality

quencies appear in the spectral envelope as the long train of side lobes. If the pulse is not rectangular, the shape of its spectrum will be different from that shown in Figure 5-7. For example, if the tuning fork is turned on and off with sloppy rather than sharp attacks and releases, there will be fewer high-frequency components and the resulting spectrum will have reduced side lobes. In the extreme case where the pulse of tone has a bell shape, the frequency spectrum has a single lobe that is also bell-shaped; the side lobes disappear.

Regardless of the detailed shape, the reciprocality relationship still holds: if you define the width of the pulse in some reasonable way, its spectrum will have a width approximately equal to the reciprocal of the pulse width.

Figure 5-11 summarizes the reciprocal properties of time and frequency that we have just been discussing. Figures 5-11a, b, and c describe the general properties of the spectra of waveforms independent of the detailed shape of the spectra. Figures 5-11d, e, and f show the dependence of the shape of the spectrum on the steepness of the waveform. This figure can serve as a useful reference when we discuss waveforms of various kinds in later chapters.

Baseband Waveforms

In the examples we have been discussing, the time and frequency properties were all based upon a tuning fork that was modulated in different ways. But all these properties are independent of the underlying tuning-fork frequency. The frequency of 100 Hz was used in the examples simply because 100 is a convenient round number. It could have been any frequency at all.

And it can also be zero; when it is, the waveform is called a *baseband waveform*. You can't buy a zero-frequency tuning fork, so the tuning-fork analogy breaks down here. But everything that we have said applies to a rectangular pulse or a repetitive sequence of such pulses with "nothing" inside. In fact, we inferred the harmonic structure of the pulsed tuning fork from the harmonic structure of the baseband modulating square or rectangular waves. Is this simply another mathematical artifice or is there some physical significance to baseband transmission?

Baseband pulses are, in fact, very real. These rectangular pulses representing 0s and 1s as in Figure 5-2 transport signals from one place to another within a computer. These signals are transferred at very high speeds with time scales quite different from our audio examples. Computer pulses are usually measured in small fractions of microseconds, rather than in seconds or milliseconds. Thus, the bandwidths of these computer pulses are measured in tens, hundreds or even thousands of megahertz rather than in the hundreds or thousands of hertz that we have seen in the audio domain. But all the principles are the same, regardless of the change in scale. Baseband pulses are practical only for transmission over short distances. For longer distances, modulation is almost always required.

Looking back at Figure 5-7, recall that the center frequency of the main lobe of the spectral envelope is the tuning-fork frequency. If the center frequency is zero, then the lobe is centered at zero, as shown in Figure 5-12. Of course, we saw the right side of this picture before in Figure 5-4. Figure 5-12 shows the left side as well. All frequencies on the left side are negative. A negative frequency is just the same as a positive frequency except for a phase change; this won't affect any of our discussions. The important thing to remember is that the spectral width depends upon the pulse shape and not the carrier frequency upon which this pulse is superimposed. Baseband pulses with no carrier frequency at all are no exception.

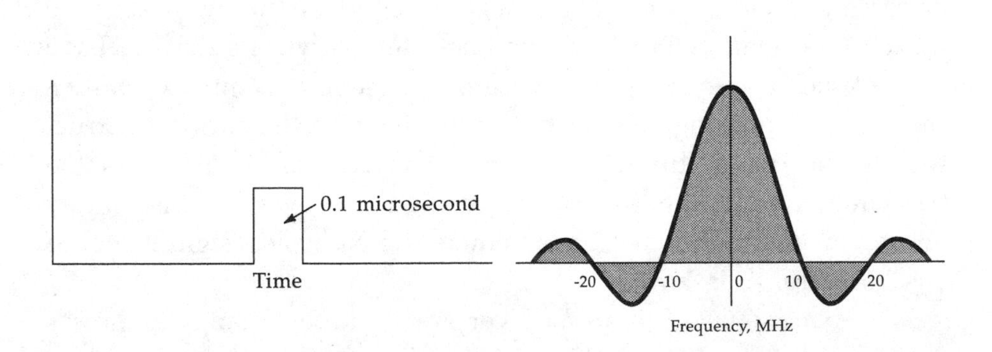

***Figure* 5-12** A Baseband Pulse and Its Spectrum

Sending Pulses over Communications Circuits

The relationships between the shapes of waveforms and their frequency content or spectrum are perfectly general. They apply to all sorts of waveforms, not simply those that are related to the pulsed tuning fork or ones like it. Knowing these properties of waveforms permits the design of waveforms suitable for transferring digital signals either within computers or between computers located at great distances from each other.

The process is conceptually simple. The basic digital communications task is to transfer a number from one place to another. It doesn't matter whether that number is a piece of data stored in a computer or a number derived from the conversion of audio sounds. This number can be thought of as a sequence of binary digits, and the first step in the process is to convert the digits into a sequence of baseband pulses. These are then moved up in frequency as the circumstance demands.

This is entirely consistent with the audio waveforms discussed in Chapter 4. The audio waveform was modulated on top of a carrier frequency appropriate to the communications medium being used. The carrier frequency for AM broadcasting is somewhere in the AM band; for FM broadcasting, the carrier is in the FM band. With a telephone line, no carrier at all is needed; the audio is sent directly.

The same thing is true with digital waveforms. The baseband signals are translated to a frequency that is appropriate to the medium being used for the communication. When the signals are being sent within the confines of a computer or even between computers that are very close together, no carrier at all is used. The baseband signals are sent directly over wires. Longer distances, however, require other techniques. If the digital signals are to be sent over radio channels, either satellite or terrestrial, they are first translated in frequency by a modulation process of some kind.

In contrast to audio transmission, digital transmission over telephone circuits does require a carrier. The reason for this, shown in Figure 5-13, is that the phone line has a passband between about 300 and 3000 Hz, while a baseband waveform is centered at 0 Hz. Therefore the pulse spectrum must be translated so that the center frequency of its main lobe is near the center of the phone line passband at 1650 Hz. Also, the width of the lobe has to be small enough so that it fits within the phone line passband. Practically speaking, the maximum data rate that a phone line can support with the binary signaling that we have been discussing is in the vicinity of 2400 bps. At this rate, most of the significant portion of the center lobe of the pulse spectrum is contained within the phone line passband. There are other modulation techniques that permit much higher signaling rates over phone lines. We will discuss these in the next chapter.

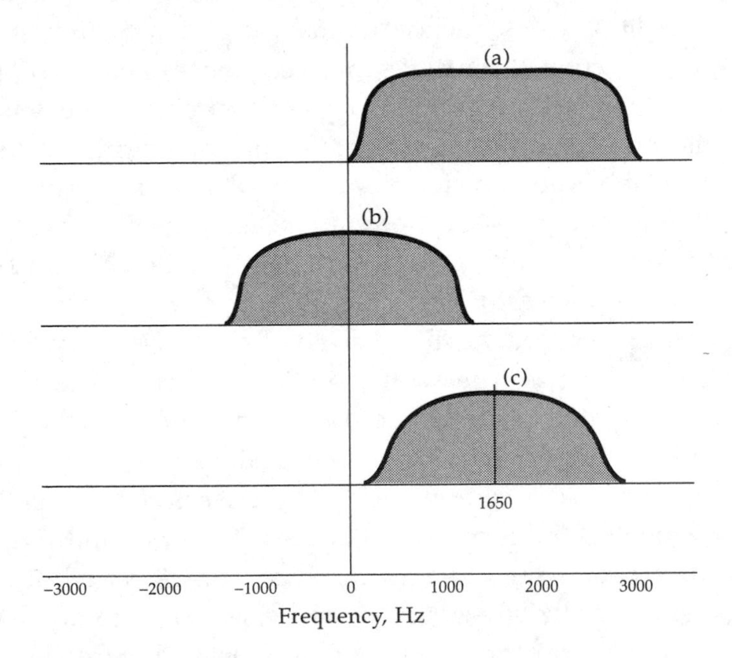

***Figure* 5-13** Sending Pulses over Telephone Lines

6

Information Theory

"The fundamental problem of communication is that of reproducing at one point either exactly or approximately a message selected at another point." Thus states Claude Shannon in the introduction to his revolutionary paper in 1948. In what he called the *Mathematical Theory of Communication*, subsequently called *Information Theory*, he went on to quantify the problem of communicating with high efficiency and accuracy, and, in so doing, converted the art of communications into a science. His approach was to address an abstraction of the practical problem faced by the communications professional on an everyday basis in having to transport his data in a *sufficiently rapid, sufficiently accurate*, and *sufficiently economical* manner.

Of course, the quantification of the adverb "sufficiently" depends upon the context of each individual application. "Sufficiently rapid" depends crucially upon the nature of the data. Live voice at one extreme must be delivered in *real-time*, i.e., as it happens with a minimal delay. At the other extreme, bulk transfer of computer files can often be delayed for overnight delivery.

Similarly "sufficiently accurate" is also dependent upon the nature of the data. When the data represent voice, music, or television, their accuracy needn't be as great as when they represent financial balances sent from a member bank to the Federal Reserve, or a set of program instructions sent from one computer to another.

What represents "sufficiently economical" is, of course, even more subjective and, as we have seen, is time-dependent, for economy has two aspects: the computational approach and the implementation technology.

Shannon's seminal papers established the performance bounds governing this three-fold tradeoff. Those who followed Shannon developed mathematical techniques for application of his theory. But widespread application of these techniques to practical problems had to await the revolution in digital component technology that led to the microprocessor and the desktop computer.

Channel Capacity

Shannon defines the *capacity* of a communications channel as the maximum amount of information that can be transported through the channel. Shannon's definition also stipulates that the information must be transferred perfectly accurately. Therefore, the capacity is the highest data rate that the channel can sustain with *perfect* accuracy. A concept defined in this way with these idealized attributes is not something that can ever be achieved practically. Nevertheless, the concept is a useful benchmark against which to compare the performance of channels when they are used with practical signaling schemes. Shannon went on to show that it is possible to achieve a balance of high data rate, high accuracy, and acceptable cost as long as the data rate is below the maximum represented by the channel capacity.

A surprising aspect of capacity is the fact that it is not zero. It is not obvious that a channel can support any transmission rate at all under the stipulation that the accuracy is perfect. Indeed, the notion of being able to transport data at an arbitrary rate and with arbitrary accuracy as long as the rate is less than some threshold value, was startling when first introduced.

Fundamental to Shannon's computation of the capacity is his concept of information—a measure of the amount of choice inherent in a family of signals. For example, a transmitted sequence consisting of three binary digits each of which can be 0 or 1 with equal likelihood, designates one of eight equally likely things and therefore carries three bits of information. But all this information reaches the recipient only if the channel supports the transmission perfectly accurately. Whenever the channel is noisy, the amount of information that is transferred is reduced commensurately. However, if the effect of the noise can be mitigated, then the *information* transfer rate can become the same as the *data* transfer rate.

Shannon's methodology in developing his theory was based upon this concept. He defined the capacity as the maximum *information* rate that the channel could support. Then he found that if the data were represented by increasingly long signals, the effect of noise in reducing the information flow became increasingly small provided that the rate at which the information was being sent did not exceed the channel capacity. The key to Shannon's methodology was the use of very long signals. Capacity itself is achieved when the signals representing the data become infinitely long. Thus, achieving capacity demands infinite complexity. The closer the communicator attempts to approach capacity, the longer the signals he must use and the more expensive his system.

Capacity, Bandwidth, and Signal-to-Noise Ratio

The two fundamental parameters that determine the properties of a communications channel are its signal-to-noise ratio and its bandwidth. Shannon's capacity must also depend upon these parameters, and the higher the signal-to-noise ratio and the wider the bandwidth, the greater the capacity of the channel ought to be.

Figure 6-1 shows the relationship between capacity and these properties of the channel. These curves plot the capacity (bits per second) against the signal power (energy per second) for a fixed noise energy for bandwidths of 1000 Hz, 3000 Hz, and 10,000 Hz. All three curves coincide at very low power levels. As the power increases, the three curves increase rapidly at first, but then diverge from one another as each flattens out at values proportional to the bandwidth. Communications engineers often refer to the left portion of the curves as the *power-limited* or *noise-limited region* and to the right portion as the

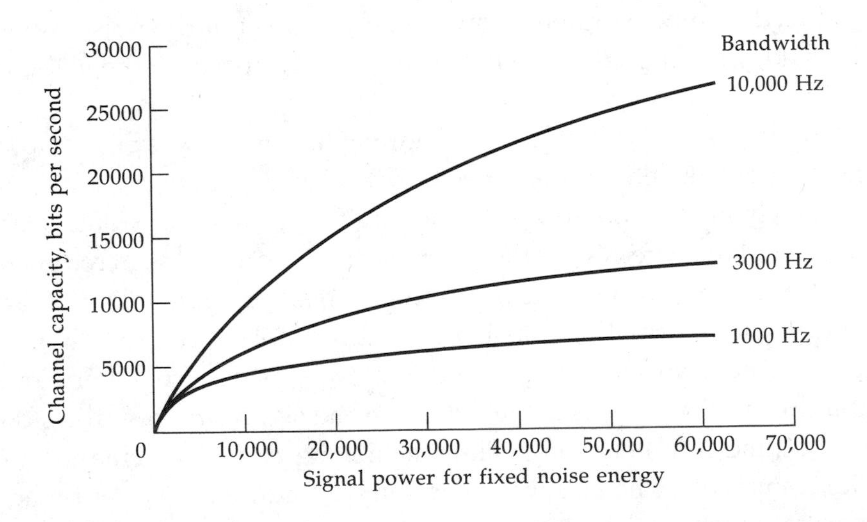

Figure 6-1 Shannon's Capacity

bandwidth-limited region. At low values of power, the data rate is confined to low values, well below the capabilities inherent in the bandwidth of the channel, just to have a high enough signal energy to maintain high accuracy. Once the power is high enough, adding more has increasingly less effect as bandwidth becomes the limiting factor.

We can gain some insight into the power and bandwidth limitations from the behavior of radio broadcast channels. In AM broadcasting, for example, a 10-kHz channel is provided to the broadcaster, and the radio transmitter must generate a signal powerful enough to provide a high enough signal-to-noise ratio that the broadcast reaches listeners within normal broadcasting range. The capacity of the channel to a receiver at a particular location depends upon the amount of the station's radiated power that reaches that receiver. Thus, at the fringes of the broadcast range, the reception sounds noisy, evidence of the power limitation that results in an inadequate signal-to-noise ratio.

Similarly, the wider the bandwidth of the channel, the more information can be transmitted. Imagine that an AM broadcaster obtained the FCC's permission to transmit in the 20-kHz band of two adjacent channels. Amplitude modulation can support the transmission of an audio signal bandwidth half as wide as the available radio bandwidth. Therefore, if the bandwidth of the AM channel were twice as wide, it could support an audio signal with twice the bandwidth, 10 kHz instead of 5 kHz. The added bandwidth would result in added signal fidelity, a form of information. But what if we tried to double the information content by doubling the bandwidth without transmitting more power to strengthen the signal? Listeners close to the transmitting station have a very-high-capacity, bandwidth-limited channel that could easily benefit from the increased information flow. In contrast, listeners on the fringe who are power-limited would find that the reception was worse because the attempted information rate was above the channel capacity. This example shows that a communications channel not only must have enough bandwidth to support the transmission of information, but also must have a sufficiently high signal-to-noise ratio to render the bandwidth useful.

A frustrating aspect of Shannon's work was that he proved the existence of channel capacity without providing a methodology for achieving it. His method was to consider all possible ways to send the information using progressively longer signals and then to compute the average amount of information transferred using all of these signals. Since we know that some of these schemes give very poor results, and these schemes are also included in the averaging process, therefore some schemes must exist that are better than average. But how do we find them? Unfortunately, the general problem of finding signals that permit communication close to capacity has resisted solution. However, we are able to do this in the power-limited extreme, and it is illustrative to examine this special case.

Families of Digital Signals

What do we mean by a family of signals? In the simplest case of binary signaling, the family consists of two waveforms, one to represent the digit 1 and the other to represent a 0. At any given time, one of these waveforms is transmitted depending upon which digit is to be sent. The process of associating waveforms with digits at a transmitter is called *digital modulation*. The signal that is finally received at the destination is a corrupted version of the transmitted waveform. The receiving processor knows precisely what the uncorrupted waveforms are like. Its job is to *demodulate* the received signal—i.e., to infer from the corrupted waveform received which transmitted waveform was more likely to have been sent. If its inference is correct, a correct digit is passed on to the user. If its inference is incorrect, an error is made. The ratio of digits in error to the total number of transmitted digits is the measure of accuracy. It is usually referred to as the *error probability*, the *error rate*, or sometimes the *bit error rate (BER)*.

Most communications are two-way. Because of this, each end of a communications link requires a modulator to send and a demodulator to receive. For practical reasons, the modulator and demodulator are built into a single piece of equipment called a modem.

Since an error is made whenever the demodulator misidentifies the waveform received, one way to reduce the possibility of confusion is to make the waveforms as different as possible. There are many ways to do this, and three of the most common are shown in Figure 6-2. In each case, the waveform is T seconds long, so that the data rate is $1/T$ bits per second. The first waveform (Figure 6-2a) is the one introduced earlier: a 1 is represented by a pulse of carrier and a 0 is represented by no pulse. This is simply a kind of amplitude modulation in which only two amplitudes are allowed. By examining the amplitude of the received waveform that has been corrupted by noise, the demodulator's job is to infer whether the transmitted amplitude was 1 or 0. The second waveform pair (Figure 6-2b) is the digital equivalent of frequency modulation. The pair is a constant-amplitude pulse of one of two carrier frequencies. In this case, the demodulator's function is to examine a band of frequencies $1/T$ hertz wide centered on each of the two possible carriers and infer which one of the frequencies was sent. The third example (Figure 6-2c), called *phase modulation,* uses a pulse of sine wave for one digit and a pulse of cosine wave for the other. Recall that a sine wave and a cosine wave are exactly the same except for a 90 degree difference in phase. In this case, the receiver knows the two possible phases and must examine the corrupted signal to see which one was more likely to have been transmitted.

In each of the three cases, it is possible for noise to confuse the receiver. The more noise energy there is relative to signal energy, the more often this will happen. Which pair of signals is best? There is no single answer for all circumstances. It depends on a number of factors—the nature of the noise, the signal-to-noise ratio, and the bandwidth of the channel, among other things. For our discussion, we will use binary FM signaling because it is best when power limited.

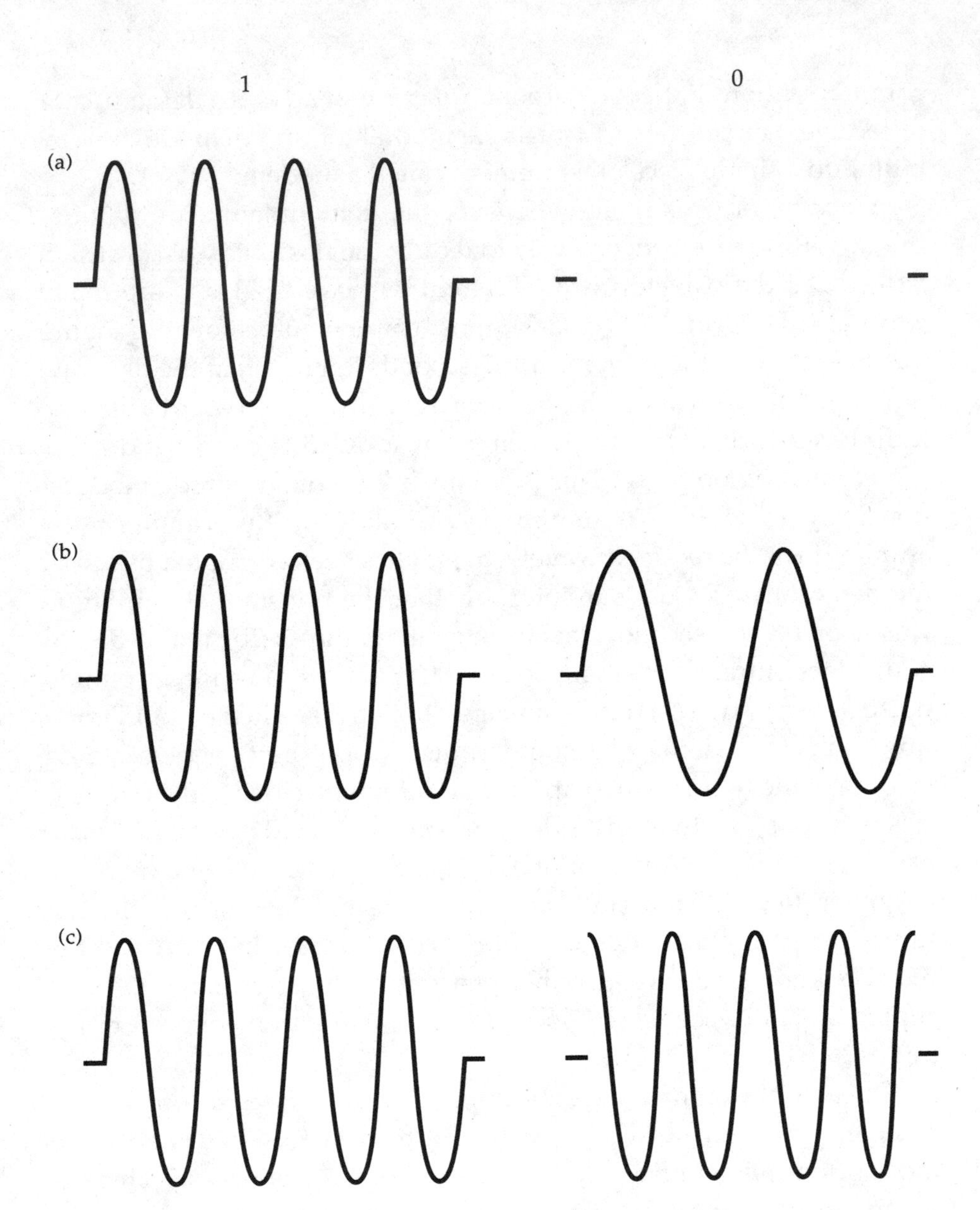

Figure 6-2 Pairs of Waveforms for Binary Signaling

Communicating with Digital FM Signals

Having chosen the signals, the first thing to be examined is how the error rate depends upon the signal-to-noise ratio. We would expect that the higher the signal-to-noise ratio, the lower the error rate. This expectation is confirmed in Figure 6-3 which shows the error rate plotted against the signal-to-noise ratio for these binary FM signals. The general shape of this curve is characteristic of all curves of error rate versus signal-to-noise ratio, whatever the signaling scheme. At very low values of signal-to-noise ratio, the error rate approaches 1/2, since in the extreme of no signal energy at all, the received signal

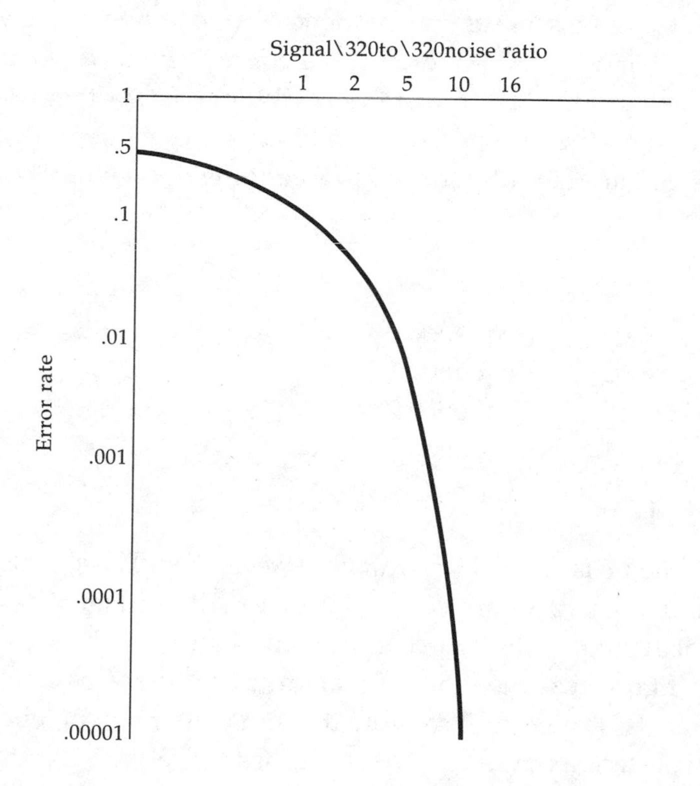

Figure 6-3 How Accuracy Depends upon Signal-to-Noise Ratio for Binary Frequency Modulation

carries no information and thus cannot shed any light on whether the transmitted digit was a 0 or a 1. Indeed, if you knew ahead of time that as many 0s as 1s were being sent, you could make your decision on each digit by flipping a coin, and on the average you would be correct half the time. As the signal energy increases, the accuracy increases, slowly at first and then more rapidly as the signal-to-noise ratio increases.

Now we introduce the issue of complexity. We saw earlier that Shannon's technique for finding the capacity was to use increasingly long families of signals, and long implies complex. Let's see what this means by generalizing the binary FM example. In the binary scheme, the transmitter accepts the bits from the data source one at a time, and for each bit it assigns one of two frequencies, depending upon whether the bit is a 0 or a 1. Now we extend the concept. Instead of accepting the bits one at a time, the transmitter will now accept them two at a time. Since there are four possible combinations of values for two bits, it will assign one of four frequencies, depending upon which combination is received:

00	frequency 0
01	frequency 1
10	frequency 2
11	frequency 3

This scheme is called *4-ary frequency modulation*. The signals for the binary and 4-ary cases are compared pictorially in Figure 6-4. If the bits are delivered by the source at the same rate in both cases, we see from the figure that the 4-ary signals must be twice as long as the binary signals. Therefore, if we hold the transmitter power (energy per second) constant, each 4-ary waveform has twice the energy of each binary waveform. But that is just what we want for a fair comparison of the two cases, since each of the double-energy signals contains two

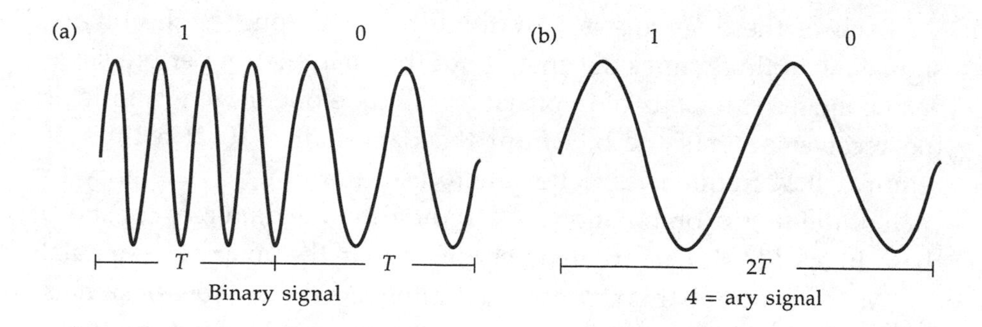

Figure 6-4 Binary and 4-ary Signals

bits of information and thus has exactly the same energy *per bit* as in the binary case. The receiver's job is now to determine from a corrupted received waveform which of the four transmitted waveforms is most likely to have been sent. When it makes this identification, it sends on to the user the two bits of data associated with the waveform. Note that the receiver's job is more complex in the 4-ary case than in the binary case, since for every decision, it must deal with four possibilities instead of two.

Now let's continue the process by taking the bits from the source three at a time instead of two at a time. By analogy with the previous example, we now need eight frequencies to correspond to the eight combinations possible with three bits. This gives us an *8-ary frequency modulation* system:

000 frequency 0
001 frequency 1
010 frequency 2
011 frequency 3
100 frequency 4
101 frequency 5
110 frequency 6
111 frequency 7

Each of these signals is now three times as long as the binary signals, with three times the energy but the *same* energy per bit, since each contains three bits of information. To take one more example: if the receiver accepts the bits from the data source 10 at a time, it requires 1024 frequencies to be able to assign a unique frequency to each of the 10-bit combinations. Again holding the data rate constant, these 10-ary FM signals are 10 times longer than the binary FM signals.

We can continue this process indefinitely. As the number of signals increases, the receiver must compare the received waveform with more possibilities. One would think that, other things being equal, the more choices, the greater the chance of making an error. But other things are not equal, since, by way of compensation, each 8-ary waveform has three times the energy of the binary waveform, and each 1024-ary waveform has 10 times the energy of the binary waveform.

Figure 6-5 shows us which scheme is best and under what circumstances it should be used. It plots the error rate against the signal-to-noise ratio per bit for the 4-ary, 8-ary, 16-ary, and 1024-ary cases along with that for the binary case, repeated from Figure 6-3. Although each of the curves shows the same qualitative behavior, there are some differences that become progressively more pronounced as the signal set becomes increasingly large. The most evident difference is that, as the signal duration gets longer, the successive curves become squarer in shape. At very low values of signal-to-noise ratio, the longer the signals, the worse the performance. However, once some threshold signal-to-noise ratio is exceeded, the error rate decreases very rapidly, and the longer the signals, the faster the rate of decrease.

This result is extremely significant. When the signal-to- noise ratio is decreased slightly in an analog system, the result is a slight degradation in quality, so slight that it may not even be noticed. A digital system is fundamentally different. A small change in signal-to-noise ratio can produce a large change in performance that becomes readily detectable. A properly designed digital system will give very good results. However, the more sophisticated the system, the squarer will

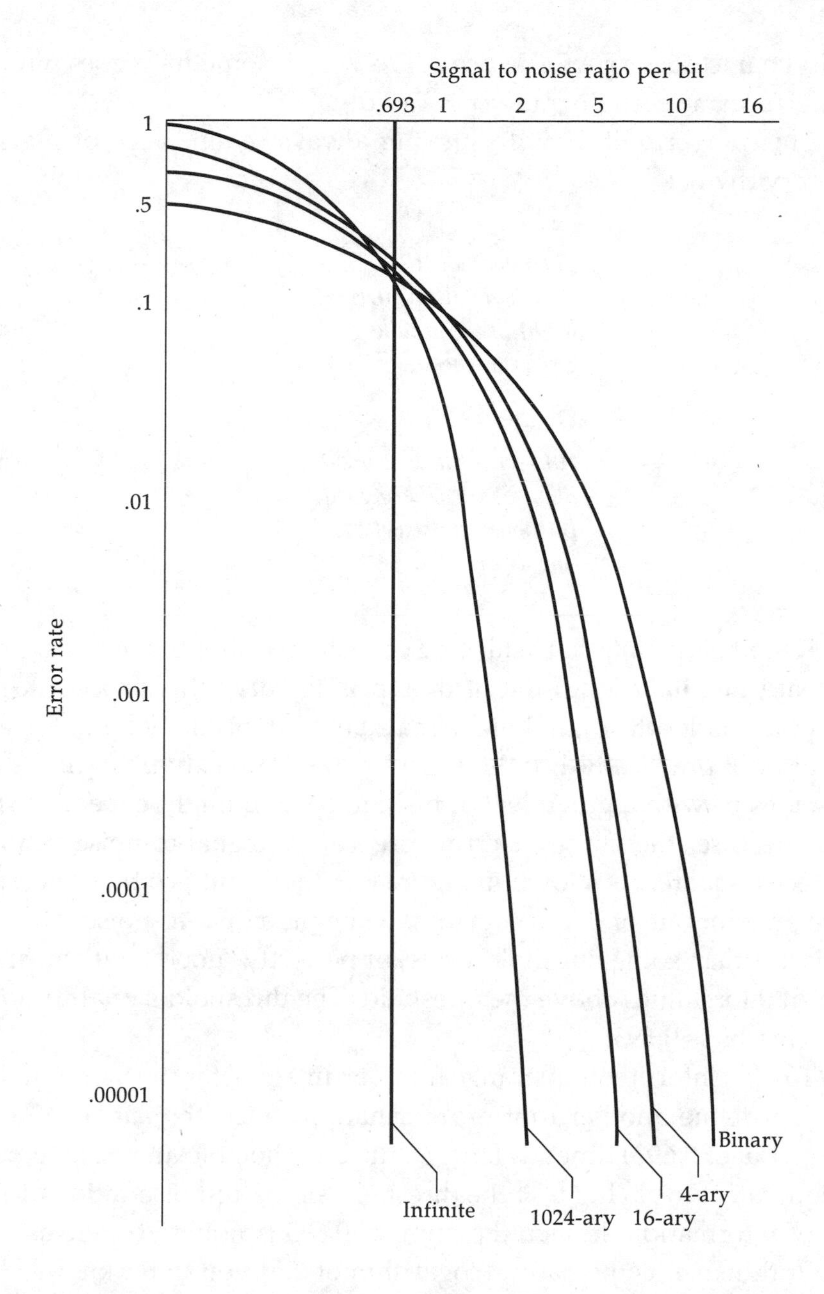

Figure 6-5 Accuracy versus Signal-to-Noise Ratio for Digital Frequency Modulation Using Different Signal Lengths

be its characteristic performance curve. And if something goes wrong, its performance can degrade very rapidly.

This property of digital signaling always reminds me of the old nursery rhyme:

There was a little girl
Who had a little curl
Right in the middle
Of her forehead.

And when she was good
She was very, very good,
But when she was bad,
She was horrid.

Keeping in mind that achieving capacity requires the use of signals that are infinitely long, the next step is to carry this process to an extreme and let the digital FM signals become infinitely long. We can never do it practically, but the consideration of this hypothetical case provides us with a great deal of insight. As you might expect, in this extreme case, the curve of error rate versus signal-to-noise ratio is perfectly square. As shown in Figure 6-5, the result is either "horrid" with an error rate of 1/2 (a coin toss) when the signal-to-noise ratio per bit is less than some threshold value, or perfectly "good" with an error rate of 0 for values above the threshold. The threshold signal-to-noise ratio per bit is 0.693.

To help interpret this strange number, imagine for the moment that we can define another unit of information called the *nat*, 1.443 (the reciprocal of 0.693) times as large as the bit. Then the above statement is equivalent to saying that the threshold signal-to-noise ratio is 1 per nat of information. In fact, the number 0.693 is not as strange as you might think—it is the natural logarithm of 2. If you don't know what the logarithm is, not to mention the natural one, just assume that the number 0.693 is one of those things like pi, a constant of nature, and skip the next paragraph. But if you do know what the logarithm is, it's worth the digression.

Binary digits of equal likelihood convey 1 bit of information. Similarly, 4-ary digits of equal likelihood carry 2 bits of information. In general, a symbol that can have M equally likely values conveys $\log_2 M$ bits of information. But we don't have to use the bit as the unit of information. We could use a decimal unit that we might call the *dit*, defined as is the bit but using logarithms to the base 10, so that the amount of information carried by a symbol with M equally-likely values is $\log_{10} M$ dits. There are $\log_2 10$, or 3.322 bits in a dit. We can, in fact, use logarithms to any base that we like. Thus, when we use natural logarithms with base $e = 2.718$, our information unit becomes the *nat* and the amount of information carried by a symbol with M values is $\log_e M$ nats. The number of nats in a bit is the natural logarithm of 2, or 0.693.

This threshold represents an absolute lower limit on the amount of signal energy relative to the noise energy needed to send a bit (or a nat) of data with perfect accuracy and infinite complexity. It is called the *Shannon limit*, and it tells us that we will never be able to communicate absolutely accurately with less energy. In fact, most practical systems use at least 10 times that much energy. Since the thermal noise energy at room temperature is about 4×10^{-21} joule, then this is the minimum amount of energy required to send a nat. The minimum amount of energy required to send a bit is 0.693 times as large or about 2.9×10^{-21} joule.

The infinitely long signals that lead to the Shannon limit also require that the channel have infinite bandwidth. We can understand this by examining the bandwidth requirements for the various digital FM signal families. In the binary case each waveform is T seconds long and it occupies a bandwidth of $1/T$ hertz centered on the frequency used for the pulse. Since two frequencies are used, one to represent a 0 and one to represent a 1, we know immediately that the channel must have a bandwidth of at least $2/T$ Hz to support communications using these waveforms. To take an example, if T is 1 ms, then each waveform will have a bandwidth of about 1000 Hz, and both together occupy 2000 Hz. The other families behave similarly:

Signaling Scheme	*Number of Waveforms*	*Bandwidth per Waveform*	*Total Bandwidth*
2-ary	2	1/T	2/T
4-ary	4	1/2T	2/T
8-ary	8	1/3T	2.67/T
16-ary	16	1/4T	4/T
1024-ary	1024	1/10T	102.4/T

Thus, increasing the number of signals from 2 to 1024 increases the bandwidth about a factor of 50. If we continued to fill in the table for longer and longer signals, the total bandwidth would become larger and larger. In the limiting case when the signals are infinitely long, an infinite bandwidth is required.

The Power-Limited Channel

Figure 6-6 shows how these digital FM signals provide the mechanism for attaining channel capacity in a highly power-limited channel. The figure repeats the lefthand portion of the capacity plots of Figure 6-1 for bandwidths of 1000 and 3000 Hz. The expansion of this portion of the curves shows how the curves coincide at low values of power, indicating the minimal influence of the bandwidth in this region. The capacity curve for a channel with infinite bandwidth is also shown in Figure 6-6. This is simply a straight line that increases indefinitely following the slope of the other capacity curves at very low power levels. This infinite-bandwidth channel is always power limited, by its very definition, regardless of the power level.

The other three curves in Figure 6-6 show the behavior of digital FM signals under the same bandwidth limitations but with an error

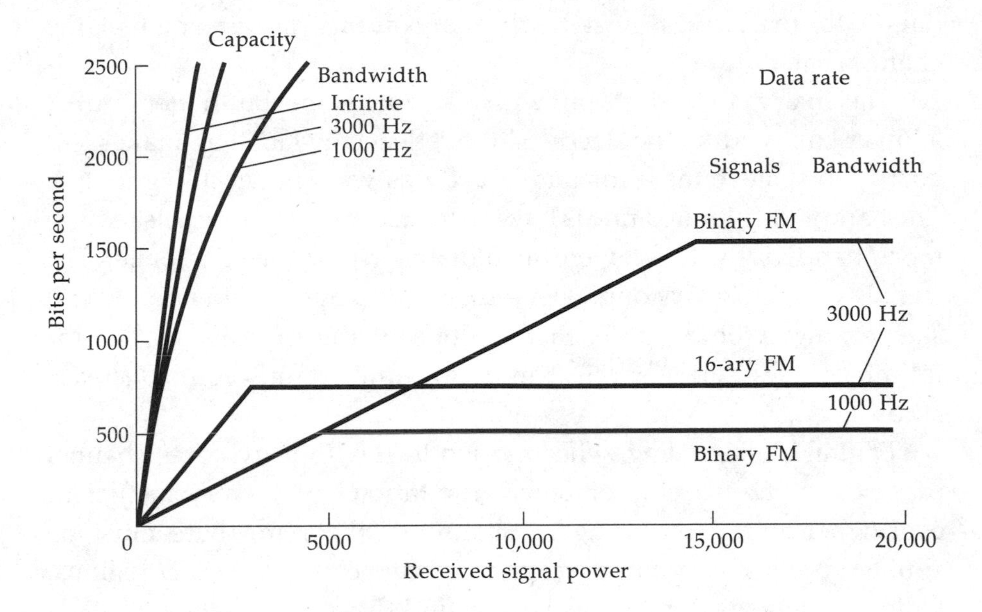

Figure 6-6 The Power-Limited Channel

rate of one in a thousand. Note that the FM curves have the same qualitative shape as the capacity curves, but are lower everywhere because nothing can exceed the capacity.

Two of the FM curves are for binary signaling with different channel bandwidths and the third is for 16-ary signaling. For these signaling schemes, the data rate is increased by decreasing the width of the pulse. This requires that the pulse power be increased proportionally to maintain the signal energy constant and thereby maintain a constant error rate. For this reason, the binary FM curves are identical straight sloping lines for low values of power. But since increasing the rate by narrowing the pulses increases their bandwidth, the rate is eventually limited by the bandwidth of the channel. Once the bandwidth of the pulse reaches that of the channel, the rate can no longer be increased. This is why the two curves flatten out abruptly. Since the bandwidth requirement for this type of signaling is twice the data rate, the curves must reach a maximum rate of one-half the channel bandwidth.

The 16-ary curve starts out with a steeper slope, but flattens out at a lower bandwidth. The steeper slope reflects the fact that it takes less energy to achieve the error rate, exactly as was shown in Figure 6-5. The bandwidth-limited rate is lower because the longer signals require more bandwidth. If we were to plot the curve for still longer FM signals—e.g., 1024-ary or greater—the initial slope would become still steeper, and would approach the slope of the capacity curves for infinitely long signals, but the bandwidth-limited rate would decrease accordingly.

Digital FM signaling, when carried to the limit, achieves channel capacity in the extreme of power limitation but is very inefficient everywhere else. According to Shannon's theorem, there must be similar families of signals that approach capacity in the other regions. Unfortunately we don't know how to find them.

Ever since Shannon's pioneering work in 1948, communications scientists and engineers have been searching for schemes that are efficient and also practical. While general results have not been achieved, over the years a set of techniques have been developed which have made such controlled-accuracy digital communications economically possible.

Practical Problems

As beautiful as Shannon's theory is, there is something unsatisfying about the way it tantalizes us with what can be done without showing us how to do it except in this one special case. The one bit of guidance that the theory provides is to use the "right" kind of long signals. And as already noted, the problem of finding such signals has two separate facets. The first is the mathematical problem of finding families of signals that achieve the desired results, and the second is the engineering problem of finding economical ways to implement these signals. Not all long signals are equally effective, and there is no general way to find signals that are optimum even if implementation cost is ignored.

Of course, complexity and cost cannot be ignored. We can gain some appreciation for the complexity problem by returning to the digital frequency-modulation example, which we know is a theoretically good approach when power limited. Suppose that, for some application, an engineer has computed that 1024-ary FM will perform acceptably. The method of generating the signals requires that the modulator accept the data 10 bits at a time and associate a waveform with each of the 1024 possible combinations of these 10 bits. The demodulator compares the received waveform with replicas of the 1024 waveforms that could have been transmitted. A number of this size is always a concern, since no matter how simple an individual comparison, a computer fast enough to perform 1024 of them might be complex. What if, later, the engineer should conclude that 10-bit signals are not long enough to achieve the required performance? Increasing the number of bits by only one doubles the number of waveforms and the number of receiver comparisons.

The term used to describe this kind of computational problem is *exponential growth*. Qualitatively, it means that a small increase in the dimensions of the problem can lead to a very large increase in complexity. Much of the research that has been done in communications has aimed to find clever ways around this problem of exponential growth.

One thing is clear: we can't transmit and receive waveforms of increasing length in the straightforward way described up to now. We must devise schemes that are clever enough to defeat the exponential problem if we are to obtain the benefits of long signals.

The first and most fundamental step is to divide the problem into two parts. First, we use a manageably small set of basic waveforms, say 2, 4, or (at most) 16. Then we generate the very long signals needed to come close to Shannon's results as *sequences* of these waveforms.

Here is an example. We are going to generate 16-ary signals. Instead of using 16 amplitudes, frequencies or phases, we will use two channel waveforms designating binary 0 and 1 and then generate 16 different sequences of these waveforms. The sequences are shown in the following list.

1.	0	0	0	0	0	0	0
2.	0	0	0	1	1	1	1
3.	0	0	1	0	1	1	0
4.	0	0	1	1	0	0	1
5.	0	1	0	0	1	0	1
6.	0	1	0	1	0	1	0
7.	0	1	1	0	0	1	1
8.	0	1	1	1	1	0	0
9.	1	0	0	0	0	1	1
10.	1	0	0	1	1	0	0
11.	1	0	1	0	1	0	1
12.	1	0	1	1	0	1	0
13.	1	1	0	0	1	1	0
14.	1	1	0	1	0	0	1
15.	1	1	1	0	0	0	0
16.	1	1	1	1	1	1	1

Each sequence contains a distinct pattern of seven 0s and 1s. The first sequence is transmitted as a succession of seven of the 0 waveforms; the second sequence as a succession of three 0 waveforms followed by four 1 waveforms etc. A comparison of these numbers gives us an initial inkling of what the process is. There are 128 possible sequences of 7-bit numbers. But since we are using only 16 of them, each sequence carries only 4 bits of information. Therefore, if we assume that 0s and 1s are equally likely, each transmitted binary digit is carrying 4/7 of a bit of information, a degree of redundancy that can potentially be exploited.

In order for sets of signals to be useful for communication, they must be different from one another. The elementary waveforms that we are using to represent the binary digits are the same as before. How about the sequences? If you examine them, you will see that every sequence differs from every other sequence in at least three places.

Suppose that the first sequence is transmitted and is received in corrupted form, so that the receiver misidentifies the fourth digit. The receiver will declare that it received the sequence 0001000. Here is where the redundancy comes in. There is no transmitted sequence corresponding to this received sequence. But the first sequence, 0000000, differs from this received sequence in only one place. If you look through the table, you will see that there is no other sequence differing from the received sequence in only one position. The next closest are sequences 4, 6, and 10, all of which differ from the received sequence in two positions. All the other sequences differ in more than two places. Since it is more likely that a single error was made in the receiving process than two or three errors, the receiver has no choice but to conclude that the first sequence was the most likely to have been transmitted.

What we have done is to divide the transmission and reception process into the two processes shown in Figure 6-7. We take four information bits and associate a sequence of seven channel bits with each information bit combination in a device called a *coder*. Each channel bit is then associated with one of two waveforms in the modulator. The received corrupted waveforms are then reconverted to channel bits in the demodulator. These bits, some of which may be in error, are then processed in a *decoder*, which will do all it can to correct the errors coming from the demodulator and then send the hopefully correct set of information bits to the recipient. By splitting the process into two, the complexity of the modulation/demodulation process is deliberately limited, and the entire scheme depends upon the complexity of the inherently digital coding/decoding process. Thus, the practical successes of the technique depend upon our ability to conceive of and implement effective coding/decoding schemes.

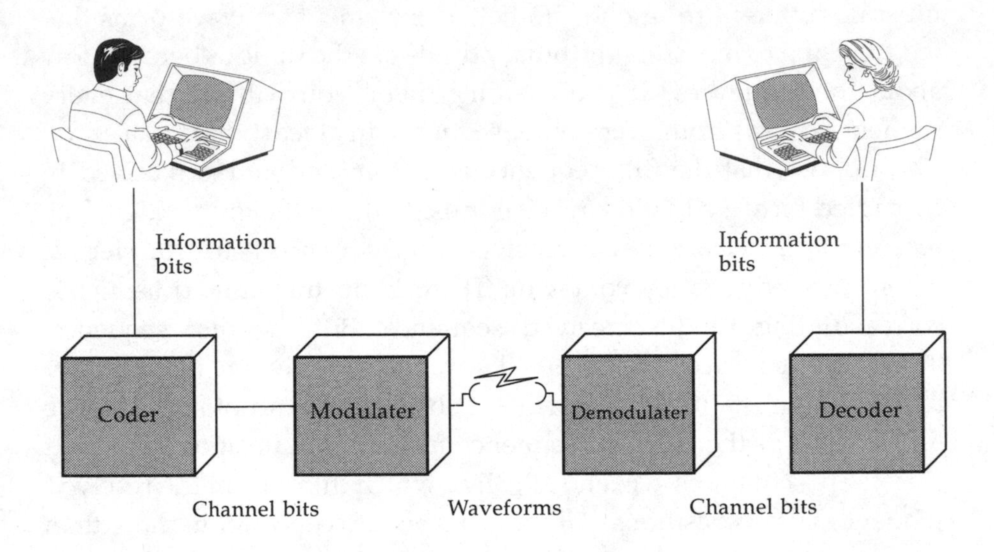

Figure 6-7 Modulation and Coding

Coding

Since Shannon's work, communications researchers have sought practical and systematic coding techniques that approach Shannon's theoretical results. The example given above is the result of work done by Richard Hamming at the Bell Laboratories, who discovered a systematic class of codes that would correct a single error. Other researchers have generalized his results, but never in a completely satisfactory way. Hamming's work and that of many of his successors is called *algebraic* coding because the decoding process takes the form of a computation in which equations are solved in a prescribed way. These algebraic approaches are useful, but they tend to deviate more and more from the theoretical results of Shannon as the codes get longer and longer. Nevertheless they are useful in many practical situations. The most widely applied of these algebraic coding classes was devised in 1960 by Irving Reed and Gustave Solomon, former colleagues of mine at the MIT Lincoln Laboratory. Curiously, its most widespread application has been to the compact disc recording and playback process, rather than to standard telecommunications. But more about that later.

There is another class of coding/decoding schemes called *probabilistic coding*. Instead of solving a set of algebraic equations to decode the received signal, this class simply takes the received corrupted signal and from it infers which of the possible transmitted signals was the most likely to have been sent. This is exactly what we did before with the digital FM signals of increasing length with which we demonstrated how to achieve capacity when power is limited. We are just obtaining the long signals in a different way. But this process, if done straightforwardly, suffers from the same defect as the earlier one. The number of computations required to compare the received signal with all the possible transmitted signals grows exponentially with the length of the signals.

But even this problem turned out to be solvable. John Wozencraft, at MIT at the time, pioneered the development of an ingenious scheme that reduces the computational complexity of probabilistic decoding sufficiently to make it practical. Wozencraft's technique, called *sequential decoding*, is a way of looking at the received digits so that they do not have to be compared with *all possible* transmitted sequences, but only with those sequences most likely to have been transmitted based upon inferences from previously decoded digits. This was the first technique that permitted communicators to make the sequences long so as to approach Shannon's results, although even this technique breaks down when the data rate is too close to capacity.

In 1961, my colleagues and I at Lincoln Laboratory built a system using Wozencraft's technique and succeeded in more than doubling the telephone line data rates customary at the time. This machine was hardly practical, nor was it meant to be. Figure 6-8 shows Professors

Figure 6-8 The First Sequential Decoder (From the Massachusetts Institute of Technology, Lincoln Laboratory)

Shannon (left) and Wozencraft (right) with Paul Rosen, the Lincoln Laboratory project manager, in front of the two large cabinets full of electronics that constituted the system. We built another system in 1965. This later version was used with one of the early satellite communications systems under power-limited conditions and was able to achieve operation at a signal-to-noise ratio of about 3, a value only 4.3 times the Shannon limit. This much smaller system took advantage of the improved component technology of the day (the first generation of small scale integrated circuits) and also used a more efficient decoding scheme devised by Robert Fano, also of MIT.

Dividing the communication problem into the two areas of modulation and coding was an essential step on the road to achieving any of the benefits of Shannon's theory. Beyond that there are no easy approaches that are applicable to all cases. However, we can simply look at these coding and modulation techniques as part of the communicator's bag of tricks to be used in individual circumstances as warranted.

But we can categorize the applications to gain some insight into the kinds of things that do make sense in practical situations. At the lowest level of sophistication, modulation techniques alone without coding may be sufficient to achieve the communicator's objectives. At the next level of sophistication, the modulation techniques alone are almost sufficient, but a coding applique is applied to reduce the error rate. At this higher level, there is no attempt to improve the efficiency of the channel by increasing the data rate. In fact, using coding in this way must decrease the data rate to obtain the benefit of improved reliability. In the earlier example of the Hamming code, 7 channel bits were used to transmit one of 16 waveforms or 4 information bits, almost 50 percent overhead. If this were applied to a modulation system for error reduction, the penalty would be almost a factor of two in data rate.

The highest level of sophistication, in the spirit of Shannon's original work, attempts to decrease the error rate *and* to increase the data rate simultaneously, or to use the metaphor of the old adage, to allow you to have your cake and eat it too. To achieve the benefits that Shannon's theory predicts, you must be prepared to apply a suitable marriage of the two techniques of modulation and coding. There is an intuitive way to see what this means. Instead of applying coding to a modulation/demodulation system that is already moderately accurate to improve the accuracy at the expense of data rate, you increase the channel data rate with the modulation system. This leads inevitably to increased errors, which are corrected with a coder/decoder of the right kind. The coding process will add overhead, but if the marriage is a good one, the increased modulation rate will more than compensate for the rate decrease in the coding process. The net result is an increased rate at an acceptable level of accuracy. The penalty in this case is simply the cost of the modulation and coding equipment.

Error Detection

There is one particularly useful special case that falls within the intermediate category cited above. Suppose you wish to transfer a data file from your personal computer to that of a colleague. The telephone line connection at the selected data rate of, say, 2400 bps yields an error rate of one in 100,000 bits, but the parameters of the application demand an error rate of no more than 1 in 100 million or more digits. You might think that, in accordance with Figure 6-3, a slight increase in signal-to-noise ratio could reduce the error rate to the desired level. But this curve is for the idealized case in which the only noise is thermal. No real system noise is ever purely thermal; there are always other perturbations. Because of this, it is very difficult, in practice, to achieve extremely low error rates in real systems by the use of modulation alone.

A very practical way to cope with this problem is to use a code exclusively for the purpose of error detection. Because the channel provides a fairly low error rate to start with, a modest code of minimal complexity can reduce the error rate to an extremely low value, and the data rate reduction due to the coding overhead is quite modest, perhaps 10 percent. These error-detection codes are algebraic, similar in principle to the Hamming code discussed above. They will detect (but not correct) any errors that may be introduced. To illustrate this point, let us suppose that two possible transmitted sequences are 0000 and 1001. If the received sequence is 1000, there is an equal probability that it results from the first sequence with an error in the first digit or the second sequence with an error in the fourth digit. Therefore there is enough information to detect the error but not correct it. Some codes combine both capabilities by correcting some errors and detecting others.

A simple example of an error-detection scheme uses principles similar to an arithmetic scheme called *casting-out nines*. Suppose you are to add a column of numbers:

$$\begin{array}{r} 1 \\ 21 \\ 76 \\ \underline{148} \\ 246 \end{array}$$

Casting out nines is a technique for checking the accuracy of your addition. It goes as follows: add together all the individual digits of the numbers, but whenever the sum exceeds 9, subtract the 9 and keep the remainder. Thus, for the above example we add 1 + 2 + 1 + 7 + 6 + 1 + 4 + 8. The sum of the first four digits, 1 + 2 + 1 + 7 = 11. Subtracting 9 we are left with a remainder of 2. Add 2 to the next two digits, 2 + 6 + 1 = 9, leaving a remainder of 0. Finally, 0 + 4 + 8 = 12 or 3 after casting out the 9. We now do the same with the digits of the sum: 2 + 4 + 6 = 12, leaving a remainder of 3 after casting out the 9. This scheme always works: the sum of the digits in the numbers to be added after casting out nines is the same as the sum of the digits in the sum after casting out nines. (I learned this scheme in elementary school and still use it to check the additions in my checkbook.)

Now suppose that you transmit these numbers over a communications channel, along with the 3 resulting from the summing process after casting out the nines. At the receiving end, the digits are summed (casting out nines) and compared to the 3. If the sum of the received digits is anything but 3, then we know that a transmission error was made. If the result is 3, we surmise that the transmission was correct. This isn't strictly true, because there are some combinations of multiple errors that can result in a sum of 3. But because these combinations aren't very likely, our surmise is likely to be correct.

The same principle holds with binary numbers. The difference is that instead of casting out nines, we cast out twos. This gives us a very simple technique for determining whether errors were made or not, because it reduces to checking whether the number of ones in the data stream is even or odd. This scheme is not useful if the probability of multiple errors is significant.

I went through this example because it is so easy to demonstrate the principle. The schemes used these days are more sophisticated and, because of this, are capable of guaranteeing error detection with an extremely low chance of an error slipping through undetected (perhaps one in a trillion). At a transmission rate of 10,000 bps, it takes about three years to accumulate a trillion bits. Therefore, in a practical sense, this kind of coding scheme may be called "error-free."

Regardless of the particular technique that you use, what is done when transmission errors are detected? A message is sent back to the sender on the channel in the reverse direction, requesting a retransmission of any block of data in which an error was detected. The detection coders and decoders are very simple and inexpensive. In fact, many telephone line modems include this kind of error-detect-and-repeat protocol as a low cost option. Such devices can be very useful provided that errors are rare and correspondingly not too many blocks must be repeated. They are typically used when files are transferred from one personal computer to another.

Such a repetition scheme is similar to the normal protocol that we use in voice communication. If we don't understand something that was said, we ask the other person to repeat the word or phrase. Were it not for this feedback mechanism, the telephone would not be a very satisfactory communication channel. But, by the same token, an error-detect-and-repeat protocol is not at all suitable for digital voice transmission, because the repetition of blocks would be likely to interfere with the normal voice conversation protocols.

To Code or Not to Code

The case just discussed is a good example of the power of coding in one particular set of circumstances. But, more generally, there may be a direct conflict between the use of modulation and coding on the one hand and modulation alone on the other. The conflict between these two extremes may be expressed as follows: is it more economical to communicate more efficiently over a given channel by using information-theoretic techniques, or to increase the capacity of the channel and use it less efficiently, thus avoiding the complication of the information-theoretic techniques? There is no general answer to this question. But specific examples applicable to bandwidth-limited and power-limited channel categories can be used to clarify how this decision is made.

The ubiquitous telephone line is the most common example of a bandwidth-limited channel. Suppose that you are the telecommunications manager for your company. You have computers in Chicago and New York with a phone line connection permitting data exchange between them. For years, you have managed with a single telephone circuit operating at the modest data rate of 2400 bps. But now your company needs a data transmission rate twice what it currently is. Should you buy another phone line to double the capacity and continue to use both of them inefficiently at relatively low rates? Or should you invest in some coding and decoding processors that let you double the rate on your old line?

When digital transmission over telephone lines was first begun in the late 1940s, the achievable rates were at most a few hundred bits per second. By the 1950s and 1960s, the standard rate increased to 2400 bps as the result of improved modulation and demodulation techniques and by carefully selected lines with the least objectionable defects. No coding was used.

In the 1960s and 1970s, a whole body of techniques was developed that permitted the modulation and demodulation equipment to compensate for the bandwidth limitations of the telephone line channels. This work led to the achievement of rates of 4800 and even 9600 bps over many phone lines. A typical telephone line modem available today for at most a few hundred dollars incorporates one or more rates in this range, together with an error-detecting code to guarantee extremely accurate data transfer. An alternative approach is to incorporate the detect- and-repeat protocol in the computer rather than in the modem. Since the protocol is a digital process, it can be implemented in either place equally easily.

But even rates as high as 9600 bps are still a long way from the channel capacity. If the rate is to be increased, the techniques just discussed must be augmented by information- theoretic approaches. Indeed the rate can be increased by as much as a factor of two on selected lines by using a form of probabilistic coding called *trellis coding*. The rate of the modulation equipment is increased to the point where errors are made, and then the errors are corrected with the coding equipment. Systems that do this became available for general use in the late 1980s.

Since the telecommunications manager mentioned above is now using his phone line at 2400 bps, doubling the rate to 4800 bps is an economical matter, since with today's technology a 4800 bps modem costs scarcely more than a 2400 bps modem. However, if he were now operating at 9600 bps, he might be forced to buy another phone line if he needed twice the data rate.

Let's take a completely different example, this time of a power-limited channel. Put yourself in the position of a NASA manager responsible to select the communications system for an unmanned probe to explore the planet Venus. The probe is to make measurements of the Venusian environment, its climate, and its geology. All these measurements are to be converted to digital form and communicated back to Earth. The capacity of this channel is determined primarily by the amount of power that the earth receiver can collect from the probe. This, in turn, depends upon the transmitter power that can be generated on the probe and the size of the sending and receiving antennas. Because sending the probe to another planet is very expensive and the cost increase with the weight of the payload, there will be severe limitations on the equipment at the transmitting end. The obvious solution is to place most of the burden upon the earth equipment and to use a large receiving antenna. The larger the receiving antenna, the more energy it can intercept and the higher the capacity. But antenna structures that are tens to hundreds of feet in diameter with dimensions accurate to a fraction of an inch can be very expensive. While a 120-foot- diameter antenna may cost 10 times as much as a 60-foot antenna, doubling its size might cost at least 10 times as much. In this example, where capacity can be very expensive indeed, it pays to use your existing capacity efficiently.

For each antenna system, the incremental cost of increasing the data rate is small as long as the data rate is well below the capacity, but increases rapidly as the data rate approaches the capacity through the use of sophisticated modulation and coding. If the data rate absolutely must exceed the capacity of a smaller antenna system, there is no choice but to use a larger antenna system, even though it is very expensive. In this application, it is highly cost effective to use very sophisticated coding/decoding, including the sequential decoding mentioned earlier, to improve channel efficiency.

A satellite communications circuit provides a less extreme example of the same phenomenon. Commercial communications companies have not yet chosen to use sequential decoding, although with present component technology, it may well be economically feasible. What is used quite frequently is a family of modulation/coding schemes called *Viterbi decoders*, named after the inventor, Andrew Viterbi, then at MA/COM Linkabit Corporation. Like sequential decoding, this scheme is probabilistic, and its implementation is quite simple for signals that aren't too long. With the right match of modulation and coding, its performance is a factor of three or four worse than the sequential decoding performance. Even so, it requires considerably less satellite power to transmit a bit than does a similar circuit without coding, and satellite power is an expensive commodity. In this way, satellite channels have become more economical as the result of the practical application of information theory.

7

Digitizing Audio and Video

The reason for communicating or recording analog sources of information digitally rather than using the more natural analog processes is to obtain the benefits of the controlled accuracy that only digital techniques can offer. The digital process modeled in the diagram below is quite simple and straightforward:

Source ⟶ A/D converter ⟶ Channel ⟶ D/A converter ⟶ Source

The analog signal is first converted to a stream of digits in a device called an *analog to digital (A/D) converter.* These digits are transported over the communications channel in much the same way as digits obtained from a computer file. Then, at the destination the analog signal is recovered by passing the digits through a *digital to analog (D/A) converter.*

The diagram also models the information flow from source to destination. The source generates information at a certain rate, and the signal reaching the destination carries information at a generally lower rate. In a perfectly accurate system, no information is lost in the process: All the information in the source is transferred to the digital stream, and the digits are received at the destination with perfect accuracy. But achieving performance close to perfection can be expensive. It stands to reason that the higher the data rate from the analog to digital conversion, the more faithfully the digital stream represents the source but the more expensive it is to communicate with high accuracy. Thus, in any practical system, the information transfer from source to destination generally requires either some reduction of the digitization rate or the communications accuracy. Which one is preferable depends upon the situation. But any system that communicates analog information in digital form requires a trade-off between the two processes.

Digitization Techniques

There are two fundamental approaches to the problem of digitizing analog signals. The most straightforward of these is called *pulse code modulation* (PCM). It has at its root the preservation of as much of the information content of the source as possible. It ignores any special characteristics of the source, taking as its basic premise the fact that a waveform is a waveform, characterized, in accordance with Fourier's analysis, by a set of frequency components. Central to this approach is another fundamental and elegant mathematical concept called the *sampling theorem* which provides the theoretical basis for the digitization process, just as Shannon's theorem provides the theoretical basis for communication. PCM provides the standard way of digitizing a signal when the objective is high fidelity or something reasonably close.

But PCM carries with it the inefficiency inherent in ignoring the nature of the source. The inefficiency manifests itself in the fact that the digitization rate usually far exceeds the source information rate. It seems reasonable that if one is forced to reduce the digitization rate significantly below that required for high-fidelity reproduction, some knowledge of the nature of the waveform source might be of considerable benefit. After all, a facsimile waveform and a speech waveform might have the same overall bandwidth but might also have significant differences that unique digitization processes could use to advantage. This is the premise underlying the second digitization approach.

But even when this second approach is taken to reduce the digitization rate, it invariably follows an initial digitization by the first approach. Thus all digitization processes whether or not they ultimately make use of any inherent knowledge of the nature of the source begin with a PCM process.

Straightforward Digitization

Figure 7-1 shows the major elements of the standard digitization process. It begins with *sampling*. A train of narrow pulses occurring at periodic intervals samples the waveform by capturing its amplitude at those particular instants of time. The waveform is converted from a continuous analog signal (point *a*) to a series of narrow pulses with the amplitudes of the waveform at those points in time (point *b*). Even though they occur at discrete points in time, these amplitudes are as analog as the waveform that they represent. They are converted to digits by a process called *quantization*. This process measures the amplitude of each sample and then converts the amplitude value into

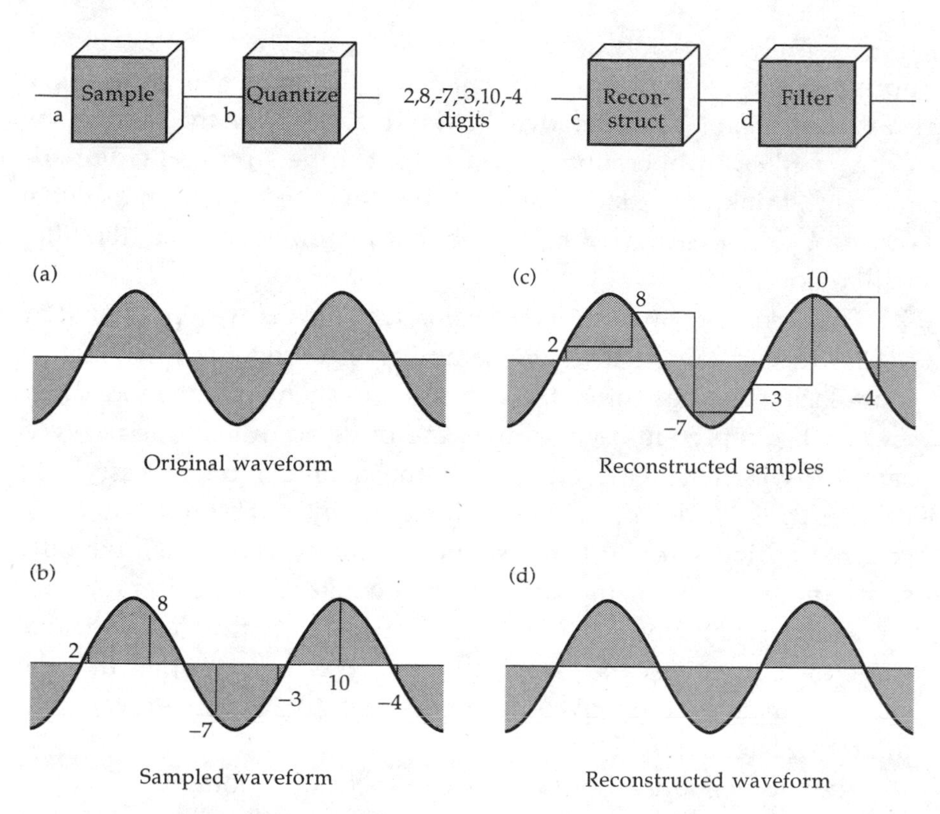

Figure 7-1 The Digitization Process

a number that has a number of digits that is consistent with the precision of the measurement. These digits are transmitted through the communications channel or stored on the compact disc or digital audio tape.

The signal reconstruction or digital-to-analog conversion process is just the reverse. First, the samples are recovered from the digits. If we assume that no errors are made in the digital processes, then the digits entering the sample recovery processing are *identical* to those leaving the quantizer. The resulting sequence of samples is almost the same as the sequence of samples entering the quantizer, differing only as a result of any inaccuracies in the quantization process. Converting these samples back to waveforms, we first obtain a crude "staircase" approximation to the original (point *c*). Finally, to recover the original waveform (point *d*), we remove the rough edges from the staircase by using a filtering operation that will be clarified presently. For the moment, think of the steep sides of the staircase waveform as being caused by the presence of high- frequency components that the filter will remove.

The accuracy with which the analog waveforms are represented by digits depends upon both the sampling rate and the quantization granularity: the more often the samples are taken and the more digits are used to represent each sample, the more accurately the received samples will match the original. If the quantization process is so precise that the difference between the quantized samples and the original analog samples is too small to be observable, then the only source of inaccuracy is the sampling process itself. The closer together the samples, the closer the staircase approximation is to the source waveform. But must the samples be so close to each other that the staircase matches the waveform almost exactly, or is a slower sampling rate adequate?

The answer to this question is of practical importance. Take the example of digital voice transmission. The total data rate needed to represent the audio signal is given by the product of the sampling rate and the quantization rate. Typically, the signal is sampled at an 8 kHz rate (one sample every 0.000125 seconds or 0.125 ms), and each of the samples is quantized into 256 levels, equivalent to 8 bits for each sample. The required channel data rate then is 8000 samples/s x 8 bits/sample, or 64 kbps. If we had to sample at twice that rate to obtain sufficient speech quality, we would require 128 kbps. Doubling the rate over a channel can be expensive whether it is achieved by doubling the capacity of the channel or by doubling the efficiency of transmission through the use of the techniques of information theory. Therefore, we must understand just what the parameters of the analog-to-digital conversion must be to understand the costs of digital transmission and, of course, recording as well.

The Sampling Theorem

The *sampling theorem* was formulated by Harry Nyquist, a research physicist at the Bell Laboratories, during the 1920s. While this is the most well-known, it is only one of Nyquist's many contributions to communication theory that paved the way for Shannon's work some two decades later. It is interesting to note that Nyquist did this pioneering work long before anyone could have dreamed how significant digital technology was going to become in the latter part of the twentieth century.

The sampling theorem addresses the problem of how rapidly samples of waveforms must be taken to retain the information in the waveforms. The result is simple and elegant. It says that if the signal contains a band of frequencies, then as long as the sampling rate is higher than *twice* the highest frequency in the band, the original waveform can be reconstructed from the samples without loss of information. To state it another way, suppose that the signal contains a band of frequency components from 0 to some maximum (the bandwidth). Then sampling the signal at a rate of twice the bandwidth or greater will permit reconstruction of the original signal with *perfect* accuracy.

But why twice the bandwidth? Why not 1.5 times or three times or ten times? The number two seems like sheer magic. In Figure 7-1 the samples are taken at a rate of five times the sine wave frequency, and it is not obvious that the crude staircase shown there can reproduce the waveform exactly.

I stated earlier that the PCM technique depends only upon Fourier's harmonic relationships. If you go back to the discussion of the pulsed tuning fork in Chapter 5 and compare the situation analyzed there with the situation represented by Figure 7-1, you can see that the same ingredients are present; only the numbers are different. Pulsing the tuning fork on and off was equivalent to multiplying the sine wave by a rectangular wave having a fundamental frequency substantially *lower* than that of the sine wave. By analogy, sampling a sine wave is equivalent to multiplying it by a rectangular wave with a fundamental frequency *higher* than that of the sine wave. The principles of Fourier analysis provide the key to the sampling theorem.

Sampling a Sine Wave

The discussion in Chapter 5 showed that signals could be represented in either the time or frequency domains. Each domain contains the same information in a different form, and sometimes the one and sometimes the other is more useful at a given time. In discussing sampling, we will lean heavily upon both representations.

We will begin with a single sine wave. Figure 7-2 shows a series of waveforms and, for each waveform, the corresponding frequency-domain picture. For example, the top picture (*a*) is of a 500-Hz sine wave. Since it is a pure tone, its frequency picture shows a single line at the sine wave frequency of 500 Hz. In (*b*), we multiply the 500-Hz sine wave by a 100-Hz sine wave. This multiplication of the two sine waves is the same as amplitude-modulating a 500-Hz carrier with a 100-Hz sine wave. This has the effect of creating two new frequencies obtained by adding the modulating frequency to and subtracting it from the carrier frequency. Therefore, the frequency picture shows two lines, one at 400 Hz and the other at 600 Hz. These lines are usually called *sidebands*—the one below the carrier is called the *lower*, and the one above the carrier is called the *upper*. You can think of the 500-Hz carrier as being split into the two sidebands by the modulation process.

Now we will do the same things for the series of sampling pulses shown in Figure 7-2c. The pulses are very narrow (0.01 ms), and they occur every 2 ms, or at a rate of 500 Hz. We have seen the spectrum of this particular carrier before. Figure 5-11 shows that the spectrum of a sequence of narrow pulses is itself a sequence of frequencies all with the same amplitude occurring at the pulse frequency and its har-

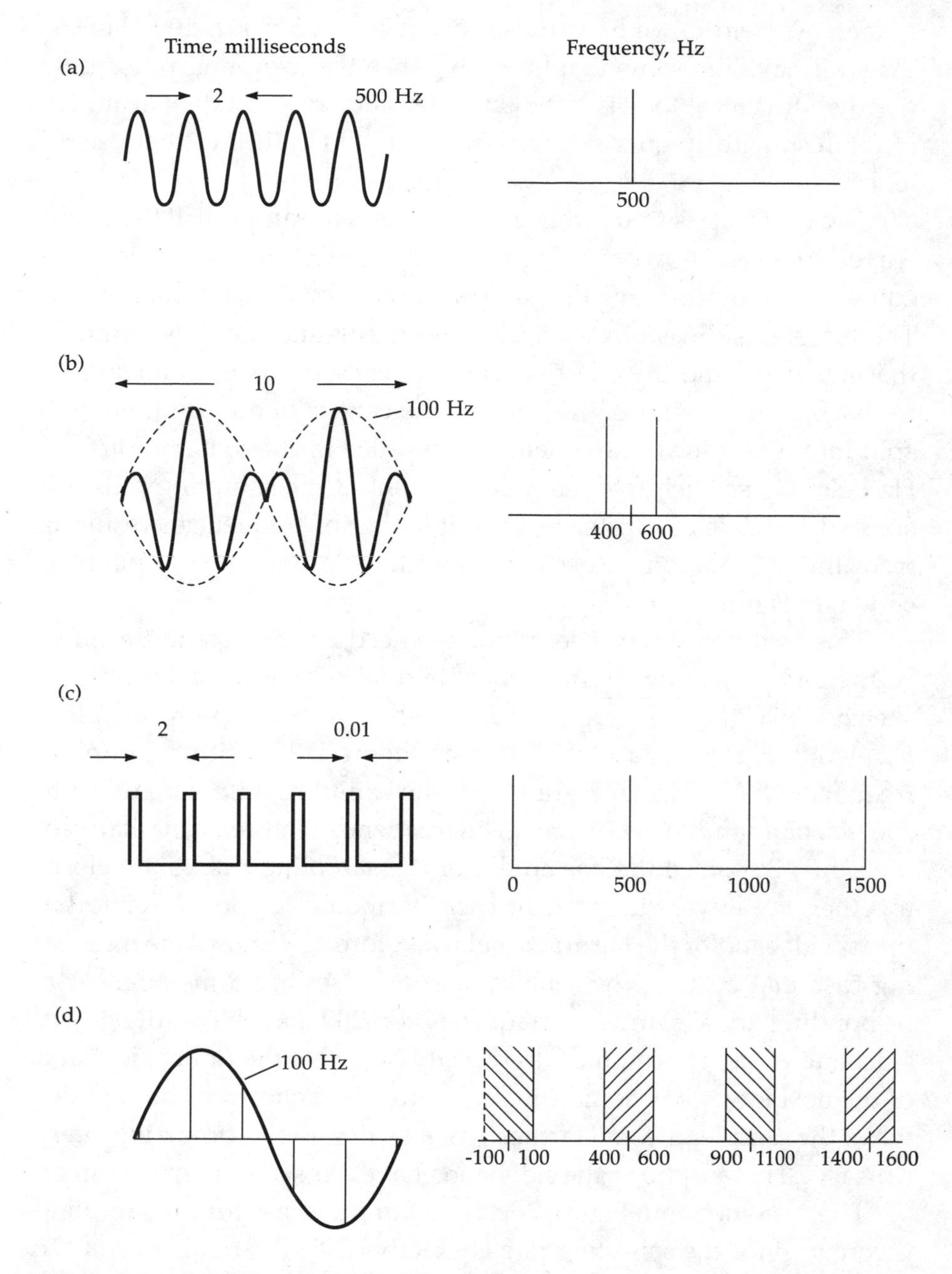

Figure 7-2 Sampling an Audio Waveform

monics. Therefore, we have pulses occurring at 500 Hz, 1000 Hz etc. They all have the same amplitude because the sampling pulse is so narrow: its central lobe is 100 kHz wide. Therefore, all the harmonics of interest, with frequencies well below 100 kHz fit into that central lobe.

The next step is to use this train of pulses to sample a 100-Hz sine wave. This is shown in Figure 7-2d. The sampling process is equivalent to multiplying the audio sine wave by the sampling pulses. Therefore, it is the same as amplitude-modulating the pulse train by the sine wave, and the effect on the frequency spectrum is analogous to that shown in Figure 7-2b. Just as the carrier frequency there was split into two sideband frequencies, one 100 Hz above it and one 100 Hz below it, so each frequency component of the sampling pulses is similarly split. Thus, instead of a single pair of frequencies resulting from the modulation process, we now have a sequence of pairs, as shown in Figure 7-2d.

This picture gives us everything we need to understand the effect of the sampling rate or frequency. Figure 7-3 shows a number of sketches like those in Figure 7-2d, each with a different sampling frequency. Figure 7-3a is the same as Figure 7-2d, with a sampling frequency of 500 Hz. In Figure 7-3b, the sampling rate is reduced to 250 Hz, or a sample every 4 ms. The frequency picture is qualitatively the same. But since the harmonics of the sampling pulses are closer together, the lower sideband of each harmonic is now closer to the upper sideband of the harmonic below. Figure 7-3c shows the interesting case that occurs when the sampling pulses are 5 ms apart, corresponding to a sampling frequency of 200 Hz. Now the lower sideband of each harmonic falls directly on top of the upper sideband of the next lower harmonic. Finally, Figure 7-3d shows what happens when the samples are still farther apart. With samples occurring every 6 ms (167-Hz sampling), the sidebands have crossed over one another.

The case shown in Figure 7-3c is the limiting case for the sampling theorem, since the sampling rate is exactly twice the frequency of the audio signal being sampled. This threshold rate is called the *Nyquist*

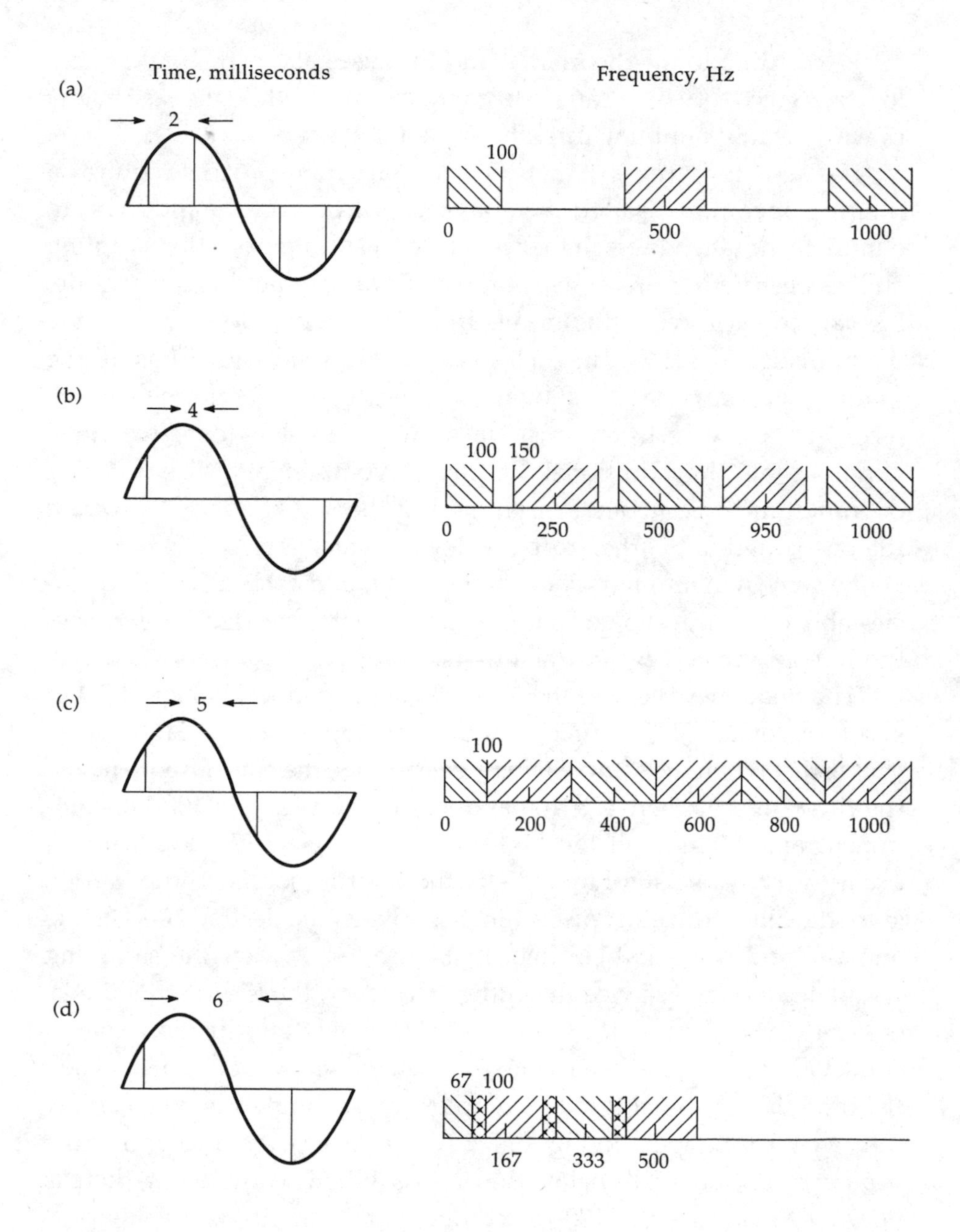

Figure 7-3 The Effect of Sampling Rate

rate. According to the theorem, it and the cases above it should result in perfect recovery of the audio from the samples, while the case below it, with the low sampling rate, should not. Let's see why.

We saw in Figure 7-1 that the reconstruction process requires a filtering operation. I stated there that the purpose of the filter was to eliminate the jaggedness in the waveform resulting from the sampling and reconstructing processes. Figure 7-1 was a time-domain picture. We can now see what this means in the frequency domain with the diagrams in Figure 7-3. In each case, the sampled signal has all the frequency components shown in the figure. But since the reconstructed waveform must be a single 100-Hz tone, we must eliminate all the additional frequency components. The filtering operation must eliminate everything but the 100-Hz line. A lowpass filter was defined earlier to be a device that passes all frequencies below some maximum frequency (the *cutoff*) and excludes all frequencies above this threshold. That is exactly what is needed here to pass the 100-Hz line and exclude all others.

The frequency pictures in Figure 7-3 are expanded in Figure 7-4 to show the effect of the filtering operation more clearly. Figure 7-4a shows that when the samples occur every 2 ms, the cutoff frequency of the filter must lie between 100 and 400 Hz to pass the 100-Hz component and exclude all the higher components. When the samples occur every 4 ms as in Figure 7-4b, the filtering job is a little harder, since the filter now must discriminate between the desired 100-Hz line and the undesired 150-Hz line. In Figure 7-4c, when the sampling frequency is exactly twice the audio frequency, the filtering job looks easier than in the previous case, but that is just an illusion. We will come back to this case in a moment, when we see what happens when the modulating signal is more complex than a single sine wave. Now observe what happens in Figure 7-4d. Since the lower sideband of the 125-Hz component falls below the desired 100-Hz component, there is no way to capture the 100-Hz component by itself with a lowpass filter. The result is the sum of a 100-Hz and 25-Hz sine wave, an obvious distortion of the original audio signal.

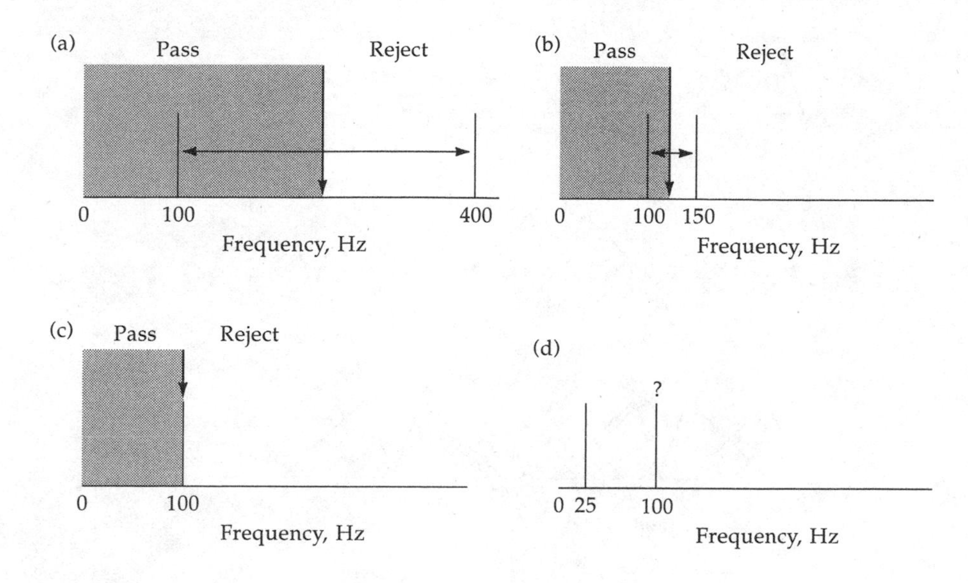

Figure 7-4 Filtering the Sampled Sine Wave

Sampling More General Signals

It is the rare exception when a source contains a single frequency. Usually, the source being sampled contains not just one, but a mix of frequencies extending over some bandwidth. The sampling process for this realistic spectrum follows directly from the single sine wave picture. Figure 7-5a shows a representative audio spectrum extending from 0 Hz up to a maximum value of 20 kHz. This might represent the spectrum produced by an orchestra playing a Brahms symphony. When a time waveform having this complex spectrum is sampled, each frequency component behaves exactly the same as the sine wave in the previous section. Therefore, when the waveform is sampled, its entire spectrum is added to and subtracted from all the harmonics of the sampling frequency. The spectrum appears as sidebands above and below these harmonics, exactly as did the line spectrum of the audio tone above. Figure 7-5b shows this for a sampling frequency of

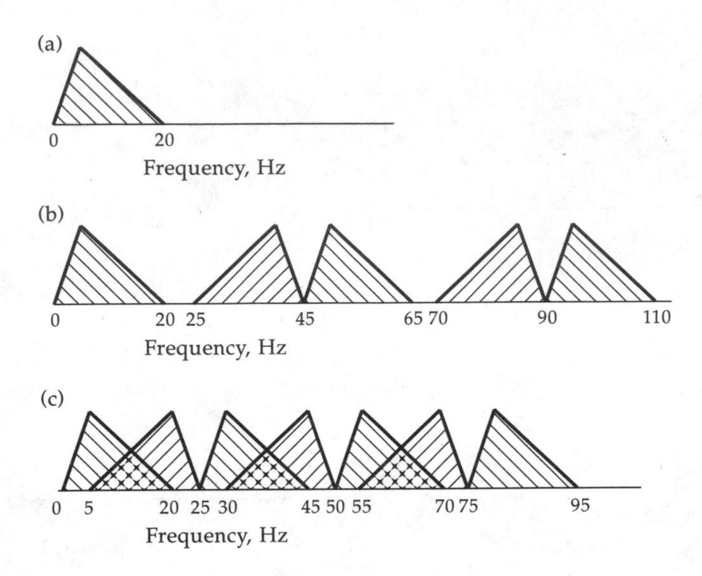

Figure 7-5 Sampling a Complex Audio Waveform

45 kHz. Since the sampling frequency is higher than twice the highest frequency in the waveform, the various copies of the spectrum above and below each sampling frequency harmonic do not overlap one another. Accordingly, a low-pass filter with a cutoff frequency between 20 and 25 kHz will allow the original signal to be recovered without ambiguity. This scheme will work as long as the sampling frequency is above the Nyquist rate of 40 kHz. When the sampling frequency is exactly 40 kHz, the spectral sidebands just touch one another. In that case, the filter cutoff frequency must be set at precisely 20 kHz to pass the entire band and eliminate the lower sideband of the first harmonic.

Figure 7-5c shows what happens when the sampling frequency is at 25 kHz, a value below the Nyquist rate. Just as in the sine wave case, the lower sideband spectra of each sampling frequency harmonic overlap the upper sideband spectra of the next lower harmonic. Once these spectra overlap there is no way to separate them out with a lowpass filter. *Aliasing* is the picturesque word used to describe what happens when the spectra overlap as the result of sampling at too low a frequency. For example, with the 25-kHz sampling frequency in Figure 7-5c, 5-kHz components are aliased as 20-kHz components, 10-kHz components are aliased as 15-kHz components, etc.

These pictures capture the essence of the sampling theorem: the Nyquist rate is the lowest possible sampling frequency that allows one to recover the analog information completely and unambiguously. However, many aspects of the pictures are idealized for simplicity. When realism is introduced, we will see that the Nyquist rate is one of those mathematical limits that can be approached but not reached in practice.

Practical Sampling Rates

Aliasing represents a gross distortion of the signal and is to be avoided at all costs. Therefore, the sampling frequency must *always* exceed the Nyquist rate. But the Nyquist rate depends on the sampled signal having a well-defined bandwidth, and signals hardly ever come that way naturally. More often than not, the signal spectrum trails off in frequency, rather than falling off sharply. When this happens, there are two approaches to avoid the sin of aliasing: (1) sample at a high enough rate so that the signal energy at half the sampling frequency is imperceptible; or (2) prefilter the signal before sampling it to artificially reduce its bandwidth, thereby permitting sampling at a lower frequency. In the first approach, all of the signal is captured with a high sampling frequency. In the second, the highest frequencies are lost, but the economy of a lower sampling rate is gained. This choice represents a compromise between bandwidth (and, hence, fidelity) and digitization rate. The wider the bandwidth, the higher the fidelity of the signal, but the more bits required to represent the signal. In digital recording, the more bits used to represent the audio, the shorter the playing time on a disc of a given size. If the choice is to limit the bit rate by limiting the sampling rate, it is always necessary to limit the bandwidth by prefiltering rather than by sampling at a rate below the Nyquist rate and thereby causing aliasing.

The filtering operation itself is also idealized. It is not possible to build a perfect filter that passes *all* the frequencies below a cutoff frequency and *none* above the cutoff. In practice, filters have a small slope-off region as shown in Figure 7-6. This means that the sampling rate must be enough higher than the Nyquist rate to allow the practical filter to discriminate between the desired and undesired frequency components. Figure 7-6b shows that, to avoid even the smallest amount of aliasing, a practical filter must use a slightly higher sampling frequency than would be needed with the perfect filter shown in Figure 7-6a.

An example illustrates the practical issues. The standard used by telephone companies for digital transmission of voice is either 56 or 64 kbps, derived from a sampling frequency of 8 kHz. This means that the bandwidth of the speech signals could be, at most, 4 kHz if the filters were perfect. Practically speaking, the actual spectrum is limited to around 3.6 kHz. The human voice has frequency components well above 4 kHz. Therefore, no matter how perfect the filters, the speech signals must be prefiltered before the sampling process to eliminate some of the higher frequency components. But analog speech transmission over the phone is also limited by the inherent bandwidth of telephone lines to something in the vicinity of 2700 Hz. Therefore whether the audio is filtered at 3.6 kHz or 4 kHz is less important than the fact that either of these frequencies is well above the bandwidth cutoff of the telephone line. From this we can see that digital telephone speech is of higher quality than analog, even taking into account the practicalities of filters and sampling rates.

In compact discs, the sampling frequency standard is 44.1 kHz, which allows the use of practical filters to recover the audio samples in a bandwidth of at least 20 kHz. Since 20 kHz is about the highest frequency perceptible to the human ear, this allows enough bandwidth for very high-quality sound reproduction.

(a)

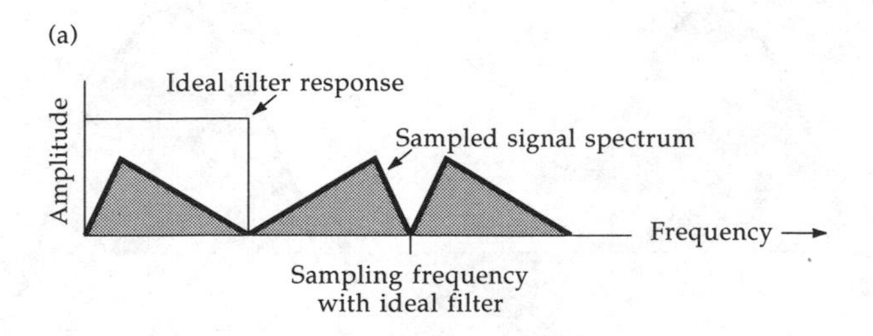

(b)

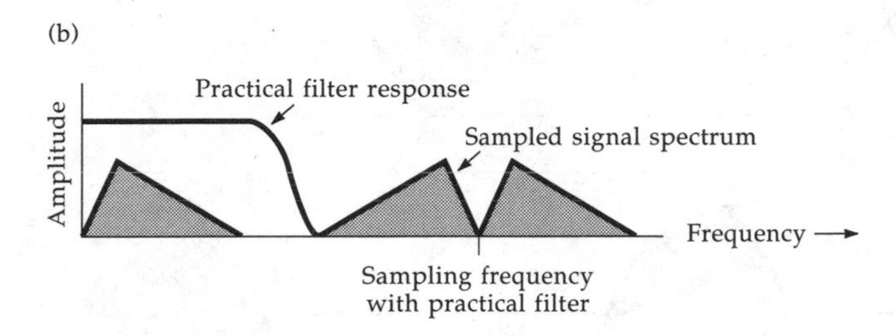

Figure 7-6 Practical Filters

Quantization of the Samples

Having sampled a waveform at a frequency above the Nyquist rate, perfect reproduction is theoretically possible with infinitely precise quantization. Let's examine the effects of realistic quantization.

Figure 7-7 shows a sine wave quantized into different numbers of levels. Each picture can be viewed as a measurement of the sample amplitude with a ruler that has a specific degree of precision. In Figure 7-7a the amplitude definition is very crude. All we are doing is measuring whether the sample is positive or negative. To do this, we need two numbers that we can call + and - or *1* and *2* or *0* and *1*, or anything else. Whatever we call them, what matters is that there are two and only two unambiguous numbers that are used to designate the sample amplitude. This is a *two-level* or, equivalently, *one-bit* or binary quantization. While it is not a very accurate way to represent

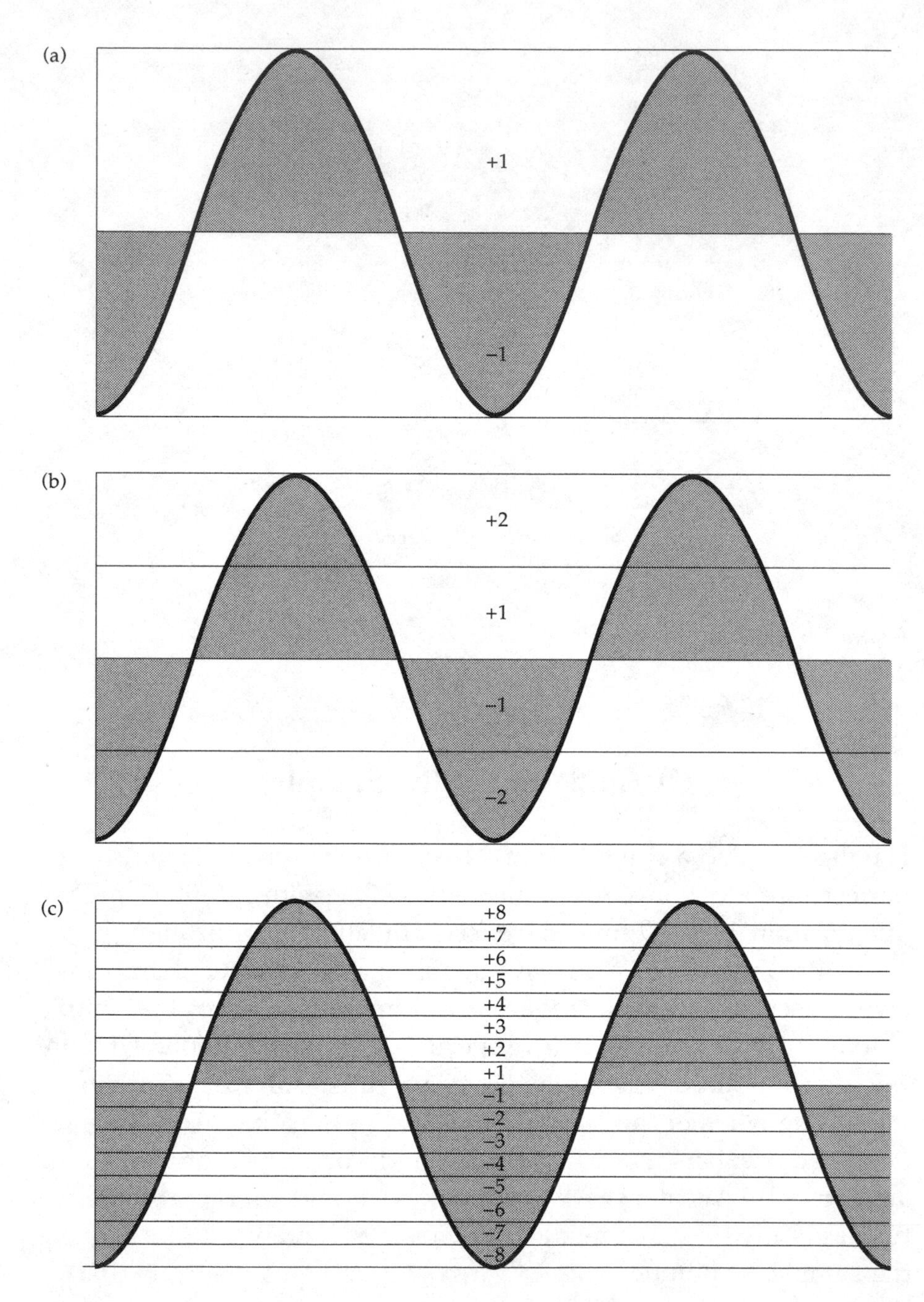

Figure 7-7 Quantization

the signal amplitude, there are cases where it is adequate. If we were to sample a speech signal and then quantize it to one bit, the resulting speech, when reconstituted in analog form, would have an unpleasant hoarse sound, but would be quite intelligible. The effect of crude quantization is similar to that of superimposing noise on top of the voice signal. For this reason, this effect is called *quantization noise.*

Four-level or *two-bit* quantization is shown in Figure 7-7b. Here we have defined four equal-size intervals, two positive and two negative. Since we can call them anything we like, let's use the designations *+1* and *+2* for the two positive intervals and *-1* and *-2* for the two negative intervals. We quantize the signal into one of these four intervals by replacing its actual amplitude with the amplitude of the interval into which the signal falls, as shown in the picture. While this representation of the sine wave is a little more accurate than that provided by the single-bit quantization, it is still quite crude. The resulting speech sounds a little less noisy than when quantized to a single bit. We can continue this process as long as we like. Figure 7-7c shows *four-bit* quantization, which quantizes the waveform amplitude into one of 16 levels, a much better approximation, but still somewhat noisy.

How much is enough? That depends on how accurate we want the process to be. Quantization to any precision introduces some level of distortion or noise; the only question is how much. With analog signals, we must control every process with a high degree of precision to preserve the integrity of waveforms. With digital signals, we determine ahead of time how accurate we want the representation to be, we sample it often enough to preserve the amount of bandwidth that we want, and we quantize it to enough levels to give us the desired level of approximation.

In all the examples shown in Figure 7-7, the intervals used to measure the signal amplitudes are all the same size. When this is the case, the quantization is said to be *linear*. The intervals do not have to be the same size. In another common quantization method, called *logarithmic*, the intervals increase in size with amplitude, thereby increasing the granularity for large samples. This turns out to be preferable to linear quantization when the number of levels must be limited for economy. This is the case for speech communication, where it is not usually practical for the number of bits to exceed seven or eight per sample. In contrast, compact discs use 16-bit or 65,536-level linear quantization. The reason for this particular number has to do with another concept called *dynamic range*.

Dynamic Range

Frequency range designates the extent of the signal bandwidth. Dynamic range refers to the range of sound loudness or, more generally, signal energy. In a system with narrow dynamic range, there is little difference between soft and loud sounds. In contrast, a system with wide dynamic range allows you to hear all sound levels from the softest to the loudest without distortion.

The dynamic range needed depends upon the context. A telephone conversation requires enough fidelity to understand what is being said and to recognize the talker, regardless of how much emotion he or she may be displaying. Requirements for music reproduction are much stricter. A recording should approximate the sound heard in the concert hall. When a great orchestra performs, the dynamic range is enormous, extending from very soft passages in which perhaps a single instrument is being played to very loud passages involving 100 or more players. Clearly, a high-fidelity recording demands considerably more dynamic range than does voice telecommunications.

The *decibel* (abbreviated *dB*) is the unit by which dynamic range is measured. The decibel is a *logarithmic unit*. One important function of a logarithmic unit is to allow a quantity with a very wide range of values to be represented with a much smaller range of numbers. Table 7-1 shows an example to help make this clear. Suppose we measure the average power in a signal using the watt as the unit of power. The

Table 7-1 Power Levels in Decibels

Power	*Decibels*
1 microwatt (10^{-6} watt)	0
10 microwatts (10^{-5} watt)	10
100 microwatts (10^{-4} watt)	20
1 milliwatt (10^{-3} watt)	30
10 milliwatts (10^{-2} watt)	40
100 milliwatts (10^{-1} watt)	50
1 watt (10^{0} watt)	60
10 watts (10^{1} watt)	70
100 watts (10^{2} watt)	80
1 kilowatt (10^{3} watt)	90

table entries show power levels increasing from one microwatt to one kilowatt by factors of 10, along with the corresponding decibel level. I have arbitrarily called the lowest power level 0 dB. For each increase in power by a factor of 10 in watts, the decibel level increases by 10. Therefore, in the bottom entry of the table, the power has increased by nine powers of 10 or 1 billion, with respect to the top entry, and the decibel level has increased correspondingly by 90. Still another way of saying this is that the largest number is nine *orders of magnitude* larger than the smallest, where the number of orders of magnitude is the number of powers of 10. The decibel is another way to represent a power of 10 or an order of magnitude.

The loudest fortissimos produced by a symphony orchestra are some 10 billion times louder than the softest pianissimos, a dynamic range of well over 100 dB. That means that a recording needs a very wide dynamic range to represent the music of a symphony orchestra with high fidelity. The typical vinyl record has a dynamic range of 60 to 70 dB, far less than the range of the music that it stores. In the recording process, dynamic compression must be introduced to keep the signal from overloading the capability of the recording medium. You may be familiar with the result. A passage of music with dynamic range so wide that it makes your spine tingle in the concert hall, ends up sounding tame in your living room. You want to turn up your volume control to approximate the concert hall effect, only to be frustrated by a combination of record scratch and amplifier distortion at the higher sound levels.

A compact disc or a digital audio tape stores signals with a dynamic range of about 90 dB. We can see where this value of 90 dB comes from by considering the implications of the 16-bit or 65,536-level quantization that is used. There are 32,768 each positive and negative zones into which the audio signals are quantized; or, in other words, the largest signal amplitude is about 32,000 times the smallest signal amplitude, or about 4.5 orders of magnitude. (Remember that 10,000 is 4 orders of magnitude and 100,000 is 5 orders of magnitude.) Also the power in a signal is the square of its amplitude. This means that if the amplitude ratio is 32,000, the power ratio is 32,000 x 32,000, or a little greater than one billion. This is just slightly greater than 9 orders of magnitude, the equivalent of 90 dB.

Decibels represent signal strength *ratios*, not absolute values of power. In Table 7-1, the lowest power level was arbitrarily called 0 dB. It could just as well have been called 20 dB, in which case the bottom entry would have been 110 dB, retaining the 90-dB difference representing the power ratio of the maximum and minimum signals. The amplifier in a hi-fi set has a dynamic range that represents the ratio of the highest to the lowest signal that it can handle. The lowest signal is determined by the noise level in the system. This includes any noise that might be introduced by the amplifier itself, together with noise that comes from the music source, the record scratch or tape hiss, or noise of any origin in the FM radio receiver. Since most modern amplifiers are virtually free of audible noise, the music source is effectively the lowest signal limit. The largest signal depends upon the characteristics of the amplifier itself. When the volume control is turned way up, the distortion of the resulting sound becomes audible. Amplifiers are rated by their maximum undistorted power output level. Therefore the dynamic range achievable in the amplifier is the ratio of this maximum power level to the noise level of the source. If the system is noisy, the dynamic range is decreased for a given amplifier power level.

A digital music source such as a compact disc or digital audio tape, in contrast to an analog record or tape source, introduces no noise at all. This fact, together with the additional dynamic range, makes digital music sources much superior to analog sources. A music system is a complex combination of many elements: speakers, amplifiers and players. In analog systems, the music source has tended to be the limiting component, and amplifiers and loudspeakers were designed to be adequate for reproducing music stored on analog tapes and records. But many of these components showed their limitations when used with compact discs. We therefore observed the phenomenon of music lovers upgrading their amplifiers and speakers so as not to limit the fundamental quality of the digital music source.

I am sure that my reaction upon hearing a compact disc recording for the first time was shared by many others. The absolute silence in the background was startling; I could hear quiet sounds that had always been at least partially masked by noise in analog recordings, and the contrasting very loud passages were breathtaking. The total experience was closer to the concert hall than I had believed possible.

Reducing the Digitization Rate

PCM is a straightforward technique designed to represent the analog signal faithfully, independent of the nature of the source and without too much regard for the resulting data rate. The channel then must pass that rate with suitably high accuracy. Might it not be preferable under some circumstances to degrade the quality of the source by digitizing it at a lower rate, either by reducing its bandwidth or by using coarser quantization, thereby easing the burden on the communications channel?

Speech

Consider the problem of transmitting digital speech. The standard digitization rate for good-quality speech is 64 kbps. If that rate could be reduced by compressing the source bandwidth in some way, then at least some money could be saved. In some cases, it might be the only way to render digital transmission feasible at all.

It is interesting to examine the relationship between the bit rate yielded by PCM and the actual information content of the signal. For example, how much information is transferred when a person talks? Let's do a simple computation. The usual speaking rate is a few words per second—-let's say 5. Assume also that a word is, on the average, 5 characters long, giving us 25 characters per second. If we use 8 bits for a character, the resulting information rate is 200 bps, a tiny fraction of the 64,000 bps that we usually use to send speech. Why the large difference?

Our calculation gave us the amount of information in the spoken words, the information that would be transferred if the speaker's words were converted to electronic mail at the speaking rate. Nothing at all is conveyed about the characteristics of the speaker's voice or of the emotion with which he or she speaks his words. The speaker might be bored, excited, or trying to end the conversation. In each case, the sound pattern would be different. Thus some of the difference between the two information rates must be related to the amount of information contained in the human voice characteristics. There have even been attempts to capitalize upon the uniqueness of the human voice by using the speech sonogram, that we saw earlier, as an identification mechanism analogous to the fingerprint. However, it is easier to mimic the human voice than fingerprints, and the courts have been reluctant to admit the sonogram as evidence.

The large disparity between the information in the text and the information in the voice speaking the text is related to the amount of choice in sending voice signals, resulting from the wide range of voice characteristics. If we could compress the speech in some way so that it is represented by a smaller number of bits per second, we would effectively reduce this choice. But at the same time, reducing the number of bits is very likely to diminish the ability of the listener to recognize the speaker and might degrade the quality of the transmitted speech in other ways.

The economic tradeoff in data rate compression is similar to the other tradeoffs that we have already discussed. By using compression, we reduce the cost of communicating the data. But this is partially offset by the cost of the compression itself. Compression makes economic sense if it costs less than the savings gained by the reduced data rate. The other cost of compression, the reduced quality of the transmission, is harder to measure in dollars and cents. Like everything else, it must be "good enough" for the application.

The simplest techniques—filtering the signal to reduce its bandwidth and sampling rate, quantizing the samples more coarsely, or some combination of the two—are satisfactory up to a point. We have already noted that the 4-kHz bandwidth used for digitization at 56 or 64 kbps cuts off the highest frequency components of some of the consonants. Therefore, the 64-kbps digital speech standard is not really high fidelity. Its quality is still quite good and more than acceptable for most circumstances, which is why it was chosen as the standard. But when we try to reduce the rate much below this value, even by as little as a factor of two, simply by reducing the bandwidth or the quantization precision, the quality begins to degrade rapidly. We need more sophisticated approaches that use our knowledge of the properties of speech.

One technique that is sometimes used to gain a factor of two in data rate takes advantage of the fact that telephone conversations are two-way—one person usually listens while the other person is talking. Therefore, although some people talk a great deal more than they listen, when averaged over (say) 10 or more speakers, you can expect that people will talk and listen about half the time. A telecommunications company will typically send a large number of telephone signals over a wideband circuit between two locations. A technique called *digital speech interpolation* is sometimes used to sense when talkers are inactive and fill in these gaps with pieces of other conversations. In this way, the overall data rate can be cut in half. Theoretically, even more compression can be gained by taking advantage of the silences between words and phrases as well. But the penalty paid for this added economy is that the low-energy beginning and end of a syllable may be clipped off; the more compression that is attempted, the more likely it is that this clipping occurs.

Digital speech interpolation uses the properties of speech conversations rather than the properties of the speech signals themselves. Another whole class of techniques does use the properties of the speech signals. The only characteristics of the source that straightforward PCM depends upon are its bandwidth and its dynamic range. But there must be other characteristics of the source that depend upon its nature. The fact that the waveform is speech must give it some properties that might help in compressing it below the rate achievable with PCM.

A key feature of PCM is that successive samples are obtained completely independently. One important class of compression techniques takes advantage of the fact that *most* of the time, successive speech samples have amplitudes that are not too different. Because of this, we can save digits by encoding successive samples relative to each other. For example, if one sample has amplitude 13 and the next sample has amplitude 15, fewer bits are needed to send a 2 (the difference between the two sample amplitudes) than to send the amplitude itself. Of course, there will be occasions when the speech amplitude changes by a large amount between successive samples. Whenever this happens, the reduced number of bits is not sufficient to represent the speech as accurately as with the full complement of bits, and there will be some degradation in quality. But the ear can tolerate this kind of degradation if it doesn't occur too often. Using this class of technique, it is relatively easy to reduce the speech rate from 64 to 32 kbps with very little degradation in quality.

There are many varieties of speech compressors in this category. The one that has become the standard in recent years is called *adaptive differential pulse code modulation*. That name is such a mouthful that its abbreviation *ADPCM* is a virtual necessity. Because of its good quality, the communications carriers sometimes use 32-kbps-ADPCM to represent speech in the digital portions of their networks where the underlying channels are wideband, sometimes in conjunction with digital speech interpolation. The cost of the compression devices is quite modest mostly because they use large-scale-integrated-circuit technology. Since the quality of speech compressed to a rate of 32 kbps remains better than that of ordinary analog telephone-quality speech, the small degradation in quality is acceptable.

How much can this class of techniques reduce the data rate? The lower the rate, the greater the likelihood that successive samples will be too far apart in amplitude to fit into the limited number of bits allotted to changes. As it turns out, the limit is about 16 kbps, while still maintaining acceptable levels of degradation. To go below this limit requires radically different approaches that make use of our knowledge of the human speech production mechanism.

These compression systems are called *vocoders*, an abbreviation for *vo*ice *coders*. They contain special-purpose computers that extract certain features of the speech based upon models of the vocal cords and vocal tract and then synthesize the speech using these features. They do this at rates as low as 2400 bps. The resulting speech at these low rates is intelligible, but of relatively low quality. Users sometimes complain that the speech sounds artificial; "Like Donald Duck" is the usual comment. In addition, when these vocoders were first built using the digital technology available at the time, they were large and expensive. Indeed they had few redeeming features aside from the fact that they constituted the *only* way of achieving security over ordinary telephone circuits.

The earliest vocoders were developed during World War II. General Eisenhower is said to have found their quality to be so objectionable that he refused to use them. As the technology advanced, vocoders were used more widely. By the mid-1960s, a vocoder had been installed on the presidential aircraft. It was used by President Johnson's aides whenever security or privacy was warranted, but not by the President himself. His aides were so fearful of his reaction to the quality of the vocoder that they declined to tell him of its installation on the aircraft. They were forced, therefore, to take the calculated risk that the President would commit some indiscretion over the air that could be potentially embarrassing. As a case in point, the story is told that when the President's airplane was returning to Washington on a route near the Canadian border, he radioed the White House to inform his staff of his seating preferences for a state dinner to be held that evening in honor of the Canadian Prime Minister. When told that his preferences would violate accepted protocol, the President is alleged to have referred to protocol and the Prime Minister in his characteristically earthy way. (We don't know whether the Canadians heard the comments or were offended by the seating arrangements. Any strain in the relations between the two countries resulting from the incident has remained a dark secret.)

There has been much progress since the 1960s in both speech research and computer technology. Today's vocoders sound quite respectable, if not high fidelity. They are about the size of a telephone console and cost about $3000.

Approaches like the vocoder are necessary to send digital speech through analog telephone circuits with bandwidths in the vicinity of 3 kHz. The highest rates supportable on the best such circuits using the most sophisticated modulation and coding techniques are in the range of 16 to 19 kbps. On the more typical dial-up lines from the telephone company, rates of 2400 or 4800 bps are usually the maximum achievable with a tolerably low error rate.

Despite all the research for so many years, we have not found a way to achieve quality comparable to analog voice with vocoder-like systems. Because of this, subscriber digital voice transmission has been used for only one application: securing speech from eavesdroppers, even though the low price of the modern instruments has allowed their use to be extended from national security applications to less critical, but still important, industrial privacy applications. This will continue as long as the analog circuit remains as the basic subscriber connection. It will only change when the telephone networks extend wideband digital service to the customer's premises and relegate the 3-kHz analog channel to oblivion.

Video

Video is another area in which compression can, in some cases, be an economical approach. Standard analog color television is broadcast in 6-Mhz channels. The actual signal bandwidth is around 4 MHz. Straightforward analog-to-digital conversion of these signals would require a data rate of hundreds of megabits per second. Much of the signal bandwidth is needed to capture the rapid motion that takes place in moving pictures. There are, however, some video applications that do not require such rapid motion. A good example is video teleconferencing—after all, how rapid can the motion be when conferees are sitting around a table? Compression devices have been developed for conference situations that drastically reduce the data rate to well under a megabit per second to create what is euphemistically called *nearly-full-motion video*. One way of doing this is to take advantage of the fact that in a conferencing situation, in contrast to a boxing match or a horse race, there is often not much change from frame to frame. Should a conferee become overly emotional and start gesticulating rapidly, his transmitted image might be blurred a bit, but that could even be an advantage.

A way to bring the transmission rate way down to a rate as low as 56 kbps is called *freeze frame television*. In this scheme, a video frame is sent once every few seconds and, in the interim, the last frame is "frozen" on the viewer's screen. I have never found this to be a very satisfactory conferencing vehicle despite its relative economy. The frozen frame makes the conference so artificial that it offers hardly more realism than a voice conference with no picture at all. Even worse, it can, on occasion, freeze an embarrassing pose on the screen.

Digital Radio

Broadcast AM radio is low fidelity. FM quality is considerably better but still well below that of the LP record. Digital radio could approach the quality of the CD. However, the amount of bandwidth needed is so great as to guarantee that broadcasting will remain analog until suitable bandwidth reduction techniques are developed. One promising experimental technique uses the psychoacoustic phenomenon called *masking*, in which the audibility of soft tones is diminished in the presence of loud tones. With this technique, the digitized audio signals are processed in a computer to remove low-energy frequency components in the presence of high-energy components, thereby reducing the bandwidth of the broadcast signals. A system called *Eureka* using this principle has been tested in France and Canada. If the experiments prove successful, these or similar techniques might become operational before the close of the century.

You Can't Cut Corners with High-Fidelity Music

But recording music is a different thing entirely. In contrast to speech, video, and broadcast music efficiency must give way to quality on all counts. The very essence of the use of digits is the fact that the digits appearing at the home playback machine are identical to those generated at the source. Therefore compression is out of the question; the original digits must be recorded. Or in the language of communications systems, the capacity of the recording medium must be large enough to accommodate the number of digits needed to represent the audio with high fidelity.

8

Digital Recording and Storage

One of the most significant aspects of the development of the digital computer has been the rapid advance in the capacity and speed of storage media. The entertainment industry has sought similar media for the storage of audio and video information, initially in analog form, but later in digital form. It would seem only natural that the same digital storage media might be suitable for both applications: after all a digit is a digit. But computer storage devices have not been economical for home audio use. However, some of the more recent developments in the digital storage of audio have begun to find application in computers as well. The compact disc is the most notable of these. But other technologies either derived from the CD or in competition with it will play similar roles in the years to come.

A History of Recording

The remarkable improvements that have resulted from digital audio are simply the latest in a long sequence of advances in audio reproduction. It all began in 1877 when Thomas Edison recorded the nursery rhyme "Mary had a little lamb" on a helical track embossed on a layer of tin foil covering a rotating cylinder. We cannot fully appreciate the benefits of digital audio without the perspective of this recording history.

This history is a tale of steady advance in the achievement of its two objectives: producing a track in a recording medium as close as possible to a replica of the acoustic signal, and playing back the recorded signal as faithfully as possible with minimum extraneous noise. So rapidly did the technology advance in the early years, and so popular did the new invention become, that by the 1890s, recordings were being made regularly, and *phonographs* (Edison's term) or *gramophones* (as they became known in Europe) became production items. By the beginning of this century, Edison's cylinder had given way to the familiar disk as the recording medium, and by 1910, virtually all the singers of the first rank were making them. We still treasure those primitive recordings made by the great singers of the era, artists such as Enrico Caruso, the Italian tenor who reigned supreme at the Metropolitan Opera during the first two decades of this century.

The invention of the phonograph occurred long before the development of electronic amplification techniques, and so the original recording and playback processes were severely constrained by the limitations of the entirely mechanical process. Vocalists were forced to sing as loudly as possible into a large acoustical recording horn that concentrated the singer's acoustic vibrations into a small volume, and this mechanically amplified sound provided the energy to cut the groove on the cylinder or record. Aside from the fact that the singers often had to strain their voices to generate enough energy, the horn introduced large distortions of the frequency components of the music. The fact that it was sensitive only to sound sources directly in front of it constituted even more serious a limitation. For this reason, those early mechanical recordings were limited to ensembles of artists small enough to be grouped directly in front of the horn, virtually excluding any but the smallest orchestral ensembles.

Playback was also mechanical. A needle riding in the groove of the record picked up the mechanical vibrations corresponding to the recorded audio. The needle also picked up unwanted scratching noises from the surface of the record. These recorded sounds were amplified to an audible level by another horn. Revolutionary though mechanical audio recording may have been in its time, its quality was nevertheless exceedingly poor by any standard. It is a pity that the great artists of the day lived before the time when recording could leave the later generations an adequate picture of their artistry.

A revolutionary change occurred in 1925, when the recording and playback processes became electromechanical. For only when the techniques of electronic amplification became available, could the recording process be freed from the limitations of the acoustic horn. The new recording device became the microphone, and the acoustic vibrations, once converted to electrical form, could now be amplified before being converted back to mechanical form for the cutting of the record grooves. Aside from the improved fidelity of the new techniques, the elimination of the horn meant that, for the first time, the size of the orchestra was no longer a limitation. Although the new process was still primitive by today's standards, its fidelity was a revelation in its day.

The next major milestone occurred in 1948 when the 78 RPM record was replaced by the *long-playing* or *LP record*. The most revolutionary change was a large increase in the storage capacity of the disk. A 12-inch 78-RPM record could store less than 5 minutes of audio on a side, necessitating a multiple-record album for a medium-length symphony less than one-half hour in duration. Recording of long works—

e.g., a full-length opera lasting two to three hours—was quite impractical. The combination of the slower rotation speed and the much narrower "microgrooves" gave the LP its "long-playing" capability, up to one-half hour per side of a 12-inch record. The LP also had improved reproduction quality, due to the development of new materials and techniques that permitted a much wider frequency response and much quieter surfaces.

Ten years later, stereo LP recordings made monaural recording obsolete. Everyone is familiar with the result. Almost every work of any length ever written has been recorded in stereo, some many times by many artists, and with increasingly high quality. In the days before stereo, any audio equipment that made even a pretense at faithful reproduction was called *high fidelity* or *hi fi* for short. However, once stereophonic recording became common, the audio equipment that reproduced it became known as *stereo,* relegating the term high fidelity to monaural systems.

Then came the incursion of digital technology into audio reproduction. The first step in the evolution from all-analog to all-digital recording was a hybrid stage in which the original audio from the microphone was recorded on digital magnetic tape. This conversion to digital at the start of the process provided a controlled-accuracy source from which the disks could be cut. The disks were still analog, and the recorded digits had to be converted back to analog before they could be cut. Some years later, the compact disc began competing with the LP as the vehicle for supplying the audio program material to the end users in digital form. I am sure that the developers of the compact disc believed that the new medium would ultimately supplant the LP. But I also conjecture that they along with everyone else were surprised at the speed at which this took place.

The compact disc is a remarkable engineering *tour de force,* a credit to those who worked so diligently and successfully upon it. While we know that Thomas Edison invented the phonograph, we don't know the name of any one individual who developed the compact disc system. It is rare these days when a single individual can be named as the father of a particular piece of technology. More often it is a large company, or, in this case, two such companies, Philips of The Netherlands and Sony of Japan, that deserve the credit.

Telecommunications and Recording: The Information Theory Connection

Engineers are strongly motivated to develop technology that will find widespread application. However they often underestimate the complexity of their technology and, accordingly, are frustrated when their inventions do not find practical application. So it was with communications engineers. Despite all the highly inventive work in applying Shannon's theory, very little was put into practice in the early years. As time went on, application in telecommunication systems increased steadily. But it took the recording industry to bring embodiments of Shannon's theories into millions of homes.

How did this come about? First, in the years between the invention and the application, the revolution in digital technology converted coders and decoders from large and expensive devices into cheap, marketable products. Second, the engineers at Sony and Philips had the insight and flexibility to make the connection between communications and digital storage that seems so obvious after the fact.

Table 8-1 shows this connection more explicitly. The compact disc is a medium for delayed communication of the digitized audio from source to destination. And the storage medium, like the communications medium, is subjected to noise processes. In the compact disc, noise arises during disc manufacture, recording, and playback and limits the equivalent storage capacity. The benchmark against which we measure the overall performance of a system is how economically we are able to approach the equivalent *capacity* of the recording/playback channel—the maximum amount of audio that can be stored and retrieved without error. More precisely, the objective is to

Table 8-1 The Telecommunications/Recording Connection

	Telecommunications	*Recording*
Channel	Transmission medium	Compact disc
Information source	Digitized audio waveform	Digitized audio waveform
Communications objective	Balance of high rate, high accuracy, acceptable cost	Balance of high storage density, high accuracy, acceptable cost
Noise	Transmission impairments	Storage defects
Channel capacity	Maximum transmission rate with perfect accuracy	Maximum storage density with perfect accuracy

obtain an appropriate balance between high recording capacity or storage density, suitably high accuracy, and economic feasibility.

The ultimate judge of audio accuracy is the human ear. The ear is relatively tolerant—the frequency of errors acceptable in an audio system is higher than the error frequency acceptable in most data transmission situations. However, that changes the problem only in degree, not in kind, since the ultimate virtue of the digital system is its ability to obtain some desired level of accuracy, whatever that might be for the problem at hand.

High recording capacity is also a very important issue. The fact that the LP record stored about 6 times as much music as its predecessor had a revolutionary impact upon the recording industry. The standard CD is 12 centimeters, or about 4¾ inches, in diameter, and has a maximum playing time of 74 minutes. An LP record is 12 inches in diameter and has a maximum playing time of slightly less than 30 minutes on each side. Thus, the CD holds about 25 percent more material on one side than does a two-sided LP. But even this modest difference can mean a great deal. There is a perhaps apocryphal story that credits the late Herbert von Karajan, then the eminent conductor of the Berlin Philharmonic, with advising Philips Corporation that a compact disc should have sufficient capacity to hold *his* recording of Beethoven's Ninth. In contrast, Karajan's older LP recording requires three sides.

The CD's relatively small size is an important feature. It means that a CD player can be small enough to be portable, or to be mounted conveniently on an automobile dashboard. It also means that less shelf space is required to store the discs themselves. But to get all that playing time on a small disc is no mean feat, and it requires the use of modern techniques based upon information theory.

Compact Discs

Capacity

A disc of a given diameter has a fixed amount of area available for use in storing audio. The smaller the area required to represent one bit of information, the more bits can be squeezed onto the disc. The bit pattern is represented on the disc as a spiral of pits and associated land areas. For playback, the digits are retrieved from the disc using a very sharply focused beam of light. The smaller the light-spot size, the smaller the pits can be and remain resolvable, up to some limit that depends upon the essential granularity of the disc material. Once that limit is reached, it does no good to decrease the light-spot size still further, just as increasing the signal-to-noise ratio in a communications channel does no good once the capabilities of the bandwidth have been exhausted. The equivalent to bandwidth in a communications channel is the number of resolvable spots on the CD's surface. The capacity of the CD as a channel is the maximum number of bits that can be stored and retrieved without error using practical laser technology.

The problem of CD recording and playback is to find a technique that achieves a packing density of bits on the disc that is a reasonable fraction of the channel capacity, with sufficiently accurate retrieval. This means using the equivalent of modulation and coding techniques to achieve performance close to that predicted by Shannon's theorem. These equivalent techniques for CD technology mean using very small pit and land areas, although necessarily larger than the minimum resolvable areas, with an error rate low enough for error-correction techniques to overcome.

Storing Signals

The mechanism for storing digits on compact discs is remarkably simple. A sequence of pits is etched in a long, spiral track that covers much of the disc area. The etching is done chemically following a process in which a laser beam draws out the intricate pattern to be etched. The disc is played by scanning another laser beam along the track. Figure 8-1 shows a sketch of a disc rotating on its turntable while being illuminated from below by the playback laser beam. Figure 8-2 shows a cross section of the disc being illuminated by this laser beam. Note in Figure 8-2 that the underside of the pits and lands is coated with a material that reflects the fine laser readout beam. The entire structure of pits and lands is only about a micrometer (40 millionths of an inch) thick. This very fragile structure is encased in a protective material slightly over a millimeter thick that provides mechanical rigidity. This material must be transparent on the bottom to permit the playback light beam to penetrate to the pits and surrounding land areas where the information is stored.

The playback process must provide a way for the light beam to distinguish between a pit and a land. Figure 8-3 shows this mechanism with an expanded view of the pit-land cross section. Note that the light reflected from the land areas travels farther than that reflected from the pit areas by twice the pit depth. This additional distance is about

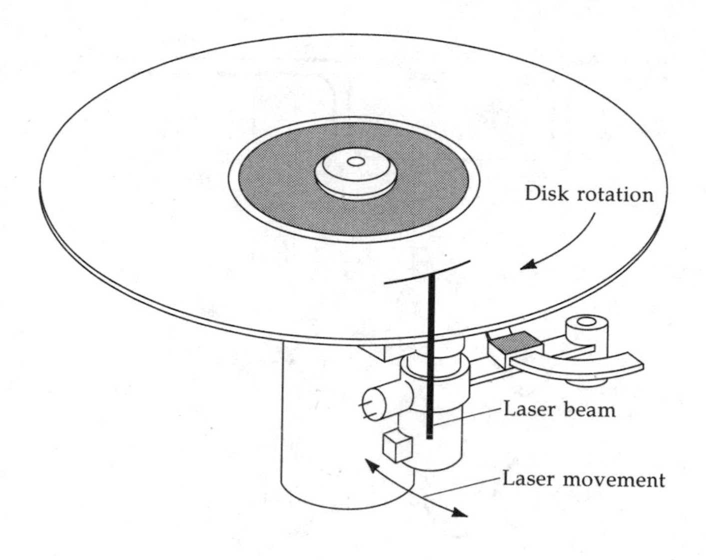

Figure 8-1 A Compact Disc Rotating above a Laser Pickup (From Richard Bruno, Making Compact Discs Interactive, *IEEE Spectrum*, 1987, © 1987 IEEE)

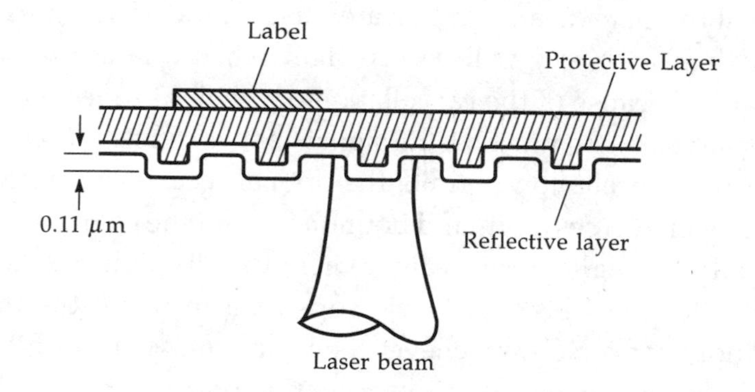

Figure 8-2 Pits and Lands on a Compact Disc (Courtesy of Sony Corporation)

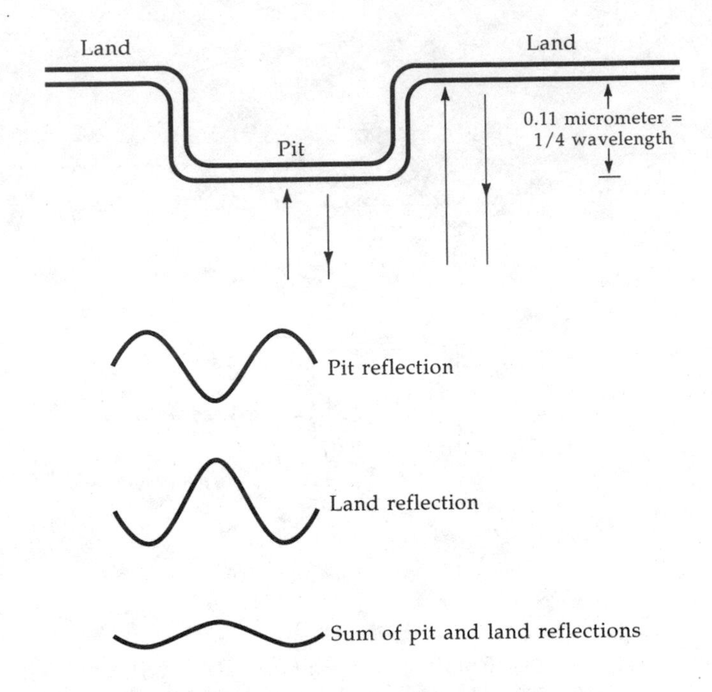

Figure 8-3 Reflections from Pits and Lands Cancel Each Other

half the wavelength of the light, causing the light reflected from pits and lands to be out of phase. The result is that whenever the beam partially illuminates a land and pit area, as illustrated in Figure 8-3, the reflected light intensity will be less than when a land area alone is illuminated, because of the cancellation of the light reflected from the pit and land areas. In practice, the signals never cancel out exactly, but even partial cancellation permits the optical receiver to detect when the light beam crosses a pit-land boundary from the changed intensity of the reflected light. Specifically, whenever the light beam crosses from land to pit, the signal level drops as a result of the increased cancellation; conversely, whenever the beam crosses from pit to land, the signal level increases as this cancellation disappears.

This phenomenon provides a convenient mechanism for storing bits. A 1 is stored at every transition between a pit and a land. The digits between these transitions are all 0s. Figure 8-4 shows a typical pattern. The optical system detects the presence of 1s by detecting the pit-land boundaries from the changes in signal level. It detects 0s by measuring the distances between the intensity changes. This technique allows the bits to be packed together closely; in fact, by a distance well under a beam width.

This scheme depends critically upon representing the digits in such a way that there are no 1s in succession. This is done with a form of code. The audio is converted to digital with 16-bit samples, and each of these samples is divided into two 8-bit segments, or bytes. These segments are the fundamental elements that are manipulated before being stored on the disc. Since the sample amplitudes can be anything at all, these 8-bit segments can contain any of the total of 256 possible combinations of 0s and 1s, many of which have contiguous 1s. The code expands each 8-bit number into a 14-bit number. It turns out that 256 of the total of 16,134 14-bit numbers (about 1.6 percent) have at least two 0s between every pair of 1s, and these are the numbers selected for the expansion. The name given to this scheme, *eight-to-fourteen modulation*, is quite descriptive if not terribly imaginative.

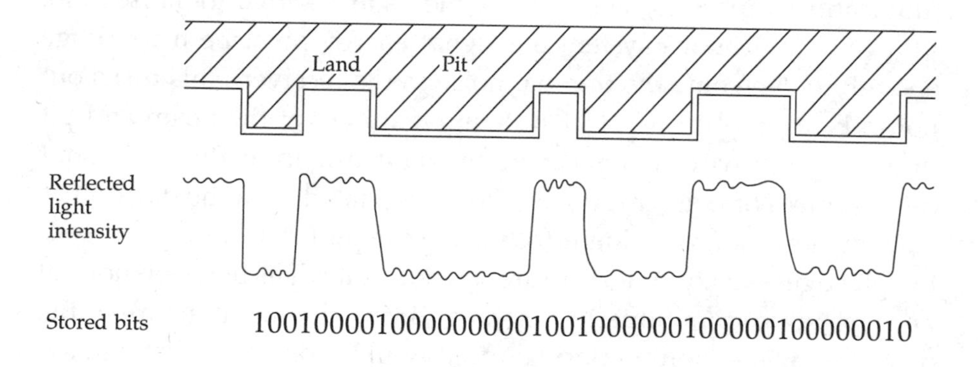

Figure 8-4 How the Compact Disc Stores Bits

An example will illustrate the process. The code transforms the 8-bit segment 01101111 into the 14-bit number 00100001000010. The original 8-bit number contains six 1s, two in a row and four in a row, while the 14-bit number contains only three 1s each surrounded by 0s. The three 1s require three transitions or 1½ pits.

What is the advantage of this scheme? For a given laser beam diameter, there is a minimum pit size, regardless of how the bits are represented. With this scheme, this minimum size pit stores three bits, since at least two 0s follow every 1. With a direct recording scheme, this same pit would be able to store only one bit. This gives us an improvement by a factor of three. When we take into account the increase in the number of bits from 8 to 14, the net improvement that results is a factor of 1.7. Other things being equal, this factor translates into the difference between a disc that can store 74 minutes of music and one that can store only 44 minutes, or, in another dimension, the difference between a 12-cm and 16-cm disc diameter.

Channel Noise

To return now to our analogy of recording and playback as communications processes, let's examine the issue of signal-to- noise ratio. In a communications system the signal energy must be made large enough to overcome the thermal noise in the receiver. This turns out not to be a problem in a CD system, regardless of the beam and pit dimensions, so we can forget about this source of noise. But how about other sources of noise, especially those associated with the disc?

One noise source is imperfections in the pit fabrication process. If the pit edge is fuzzy, or if there are spurious pit edges between normal ones resulting from imperfect etching during the manufacture of the disc, the signals representing 1s and 0s will become less distinguishable. The smaller the pit size, the more important these anomalies become; or, equivalently, the less the signal-to-noise ratio becomes.

Because these pit imperfections are random, the effect is very similar to that of thermal noise in communication systems.

Pit anomalies represent only one noise mechanism on compact discs. More significant are imperfections on the disc surface itself, which can make large sections of the disc either unusable or at least very unreliable. This results in lengthy bursts of errors similar in some respects to the large error bursts that often occur on telephone lines.

Still another source of noise is surface contamination of the disc: for example, scratches and fingermarks caused by careless handling. Fortunately, surface contamination is of little concern if minimal care is taken. The reason is that the light beam used to retrieve the data from the disc is focused to a very fine point on the pits themselves. The beam is out of focus at the surface of the disc, so any surface imperfections are smeared out by the light beam. In fact, the degree of care required is far less than that required in the handling of vinyl analog records.

In the same way that communications channel noise becomes more of a problem as the data rate increases, all three CD noise sources become more important as the pit size decreases (or as the recording density increases). The techniques used to combat this CD recording/playback noise are similar to those used in communications systems—error correction and detection.

The single most annoying source of noise in an analog recording is not present in the compact disc, because the CD playback mechanism is optical rather than mechanical. In contrast to the analog record, nothing physical touches the CD during playback. Thus as long as the surface is not physically abused in handling there is nothing in the compact disc equivalent to the noise that we call "record scratch," nor is there the corresponding record wear that ultimately degrades the quality of analog recordings. Part of the evolutionary improvement in analog techniques that occurred over the years was the development of playback systems that required increasingly smaller forces exerted on the record by the playback stylus. But no matter how light the tone-arm and how low the stylus pressure on the disk, there must be some, and with each playing of the disk, the record surface is eroded ever so slightly and eventually becomes too noisy to use. The radio stations vastly prefer CDs to LPs because of their durability, despite the fact that the improved quality of the digital technique is, for the most part, lost in the broadcasting processes.

Controlling Channel Noise

We have seen how the efficiency of a communications channel is increased by increasing the modulation rate to the point where errors are made and then applying coding techniques to correct those errors. That is precisely the philosophy that underlies the choice of parameters of the compact disc. Very small pits (about 10,000 per centimeter or 25,000 per inch) are used, resulting in an error rate due to the pit imperfections of between 1 in 10,000 to 1 in 100,000. If these pit imperfections, which behave like thermal noise, were the only consideration, it would pay to reduce the spot size still more and correct the increased error rate with still more powerful coding. However, it turns out that the limiting factor is actually the disc imperfections that make large areas of the disc relatively unreliable; and the smaller the pit size, the greater the size of the error bursts that must be corrected.

The code used to correct CD errors is one of the Reed-Solomon algebraic codes mentioned in Chapter 6. No one would have been more astonished than Reed and Solomon if someone had predicted that their decoders would be found in millions of homes 25 years after their development.

The Reed-Solomon code used appends 8 redundant bits to every 24 information bits derived from the analog-to-digital conversion process. This coding overhead of 25 percent allows almost all the random errors to be corrected, thereby reducing the random error rate to about one in a billion. Even more significant, the coding can correct large bursts of errors and detect many more without correction. Discs containing areas of imperfection so large that they cannot be handled adequately by these coding techniques are weeded out in the quality control processes at the manufacturing plants.

One other technique is used to help counteract the effect of large bursts of errors. The audio samples coming from the analog-to-digital conversion process are not stored consecutively on the disc as would be most natural. Instead, before the modulation and coding processes are applied, the samples are interspersed with samples taken at different times. This distributes the effects of a bad spot on the disc over a long period of time rather than concentrating them over a short time interval. The result is to make long bursts of noise more like random thermal noise, thereby increasing the effectiveness of the Reed-Solomon coding.

Recording

A summary of the compact disc recording processes is shown in Figure 8-5. First, audio samples are taken at the standard rate of 44,100 samples per second. These samples are quantized to 16 bits to obtain the desired dynamic range. For recording purposes, the basic time element is 135.6 microseconds, the time it takes to accumulate 6 audio samples at the sampling rate. The number of bits of audio in this basic unit is 12 samples, 6 from each of the stereo channels, times 16 bits per sample, giving a total of 192 bits. Next, the Reed-Solomon coding adds one redundant bit for every three information bits increasing the number of bits from 192 to 256. Then comes the eight-to-fourteen modulation that expands the number of bits from 256 to 448. Finally, an additional 140 bits are appended for bookkeeping purposes, giving us a grand total of 588 bits to be recorded on the disc as the basic recording unit.

We have expanded the original 192 bits of audio by a factor of about three into 588 channel bits. By doing this, we achieve high-density, accurate recording and playback according to the prin-

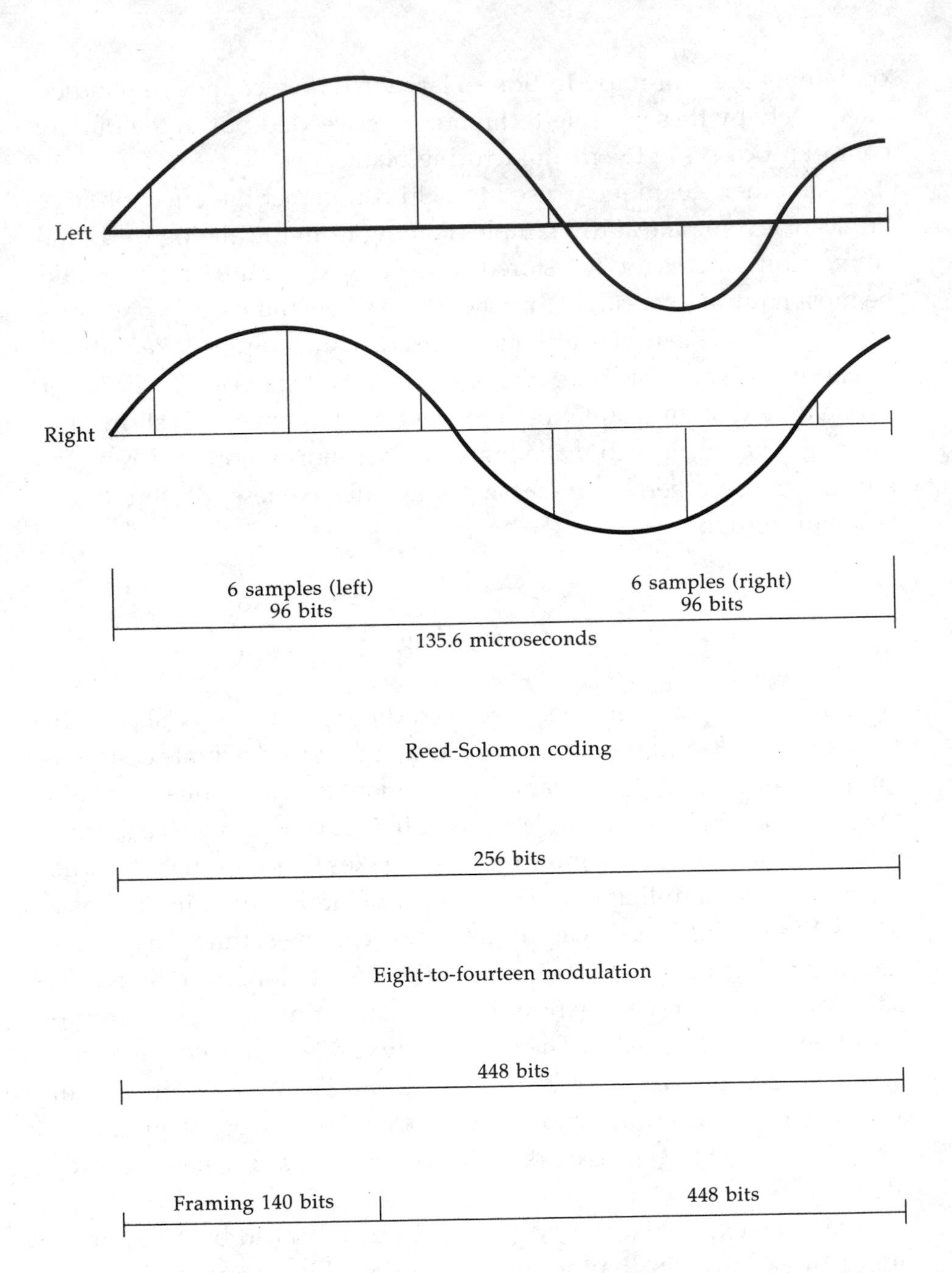

Figure 8-5 Compact Disc Recording Processes

ciples of information theory. A total of about 20 billion of these channel bits can be stored on the disc, corresponding to about 74 minutes of audio.

Playback

The playback process must undo all the operations shown in Figure 8-5 in reverse order. The optical readout system in the CD player retrieves the modulated bits stored on the disc. The optical mechanism must track the recorded pits on the spiral track with sufficient accuracy to keep this retrieval from being a source of errors. The selection of the minimum pit size is governed by both the size of the light spot and the accuracy with which the spot tracks the pits on the spiral. These bits are demodulated by reversing the eight-to-fourteen modulation and then are decoded using the Reed-Solomon decoding process. The resulting digits are ready to be converted back to audio and then played through an amplifier and speaker.

But first, some more error processing is needed. The Reed-Solomon decoding will correct some errors and will detect others that it cannot correct. What should be done about the latter? Suppose that a particular sample contains an error somewhere that cannot be corrected. If there is a large difference between the incorrect sample and the adjacent correct samples, conversion of the incorrect sample to audio could produce a very annoying "pop." Rather than reproduce the audio from a sample that is known to be in error, it is preferable to ignore the received sample altogether and set the value of that sample at a value between the preceding and following samples known to be correct. Generally, the human ear is not able to detect these interpolations provided that they occur relatively infrequently.

The Success of the CD

The success of the CD was helped by one factor that isn't given as much credit as it should be: standardization. The LP record did not replace the 78 as the analog recording standard immediately. At the same time that Columbia Records developed the 33⅓ RPM LP, RCA developed the small 45 RPM discs. While the LP ultimately won out, the two schemes existed side by side for a number of years, with the LP used primarily for the classics and other longer works, and the 45s for the popular music of the day. The same thing happened with video cassette recording. Two incompatible schemes, VHS and Beta, were developed and competed side by side for many years.

In contrast, only one CD system was ever marketed. This has meant that the manufacturers of the players were able to concentrate on mass producing a single class of products. This helped contribute to the rapid decline in the prices of the players bringing them within the range of almost everyone in our society.

We should not lose sight of the fact that, despite its spectacular quality, the CD is still a recording. As good as it is, the CD is still only an approximation of the concert hall. I jump at the opportunity to hear music in a live performance, even when performed by musicians less skilled than the many outstanding recording artists. For no matter how excellent the recorded performance, no matter how high the quality of the reproduction, every time I listen to the recording I am hearing exactly the same thing. The live performance, in contrast, may be technically less polished but, in compensation, may offer a fresh insight into the music. This is especially so with jazz, where the ingenuity and spontaneity of its performers are more important than the inventiveness of its composers, and to the music of the Baroque composers who left considerable leeway to the performing artists. But even in music where the composer has specified instructions to the performers in elaborate detail, there is ample latitude for the latter to exercise his or her own esthetic judgment. To approximate the spontaneity of the concert hall, even at the risk of deviation from technical polish, some artists (Leonard Bernstein has become a notable example) insist that their recordings be made from live performances rather than the more typical studio recording sessions.

Even aside from the spontaneity of a live performance, the sheer sound of live music in the concert hall plays a key role in the esthetic experience derived from the performance, and even the best audio system does not always capture the concert-hall vividness. I remember the first time I heard a live performance of the Brahms Requiem many years ago. In the second movement, a drum sounds a repeated pattern of progressively increasing intensity, followed immediately by the words of Isaiah "For all flesh is as grass, and all the glory of man as the flower of grass," sung by the chorus in a sustained fortissimo. I have never forgotten the impact of that low-pitched drum crescendo. No LP version even approximates the effect. Compact disc versions with their wide dynamic range are much more realistic, but even when played through a high-quality amplifier and speakers, they don't have quite the bite of the concert-hall sound.

More and more often though, sound passages like this are the exception rather than the rule, and as the quality of recordings has improved over the years, with the fidelity of the playback equipment keeping pace, the living room has become an increasingly acceptable substitute for the sound quality, if not the spontaneity of the concert hall.

LaserDiscs

Optical disc technology appeared in the seventies for the recording of video, before the CD hit the market. The principal application was home movies. Recorded on both sides of a 12-inch platter, the *Laser-Disc* could hold about 2 hours of video programming with a sound track. Despite this remarkable piece of technology, the LaserDisc made hardly a ripple. Because no sooner did it appear, then along came the video cassette recorder which offered a way not only to play pre-recorded material, but also to record TV programs off the air and to make home video recordings. The first LaserDiscs were analog, as were the competing VCRs. Therefore, LaserDiscs had no substantial quality advantage, and the optical video technique faded into a temporary oblivion.

Following the success of the CD in bringing digital audio to the market, however, the LaserDisc made its second appearance, this time with a digital sound track accompanying the still-analog video. With this modified format, the LaserDisc offers sound quality distinctly superior to that available on the VCR. It is a natural medium for high-quality presentations of operas and symphony videos where the sound quality must be first-rate. LaserDisc players now on the market permit the sound track to be played on audio equipment independent of the television receiver, and this new flexibility has increased the marketability of the optical technology.

The analog video recorded on LaserDiscs is only slightly better than video recorded magnetically. Since the optical disc can store digits that are derived from audio waveforms, it can also store digits that are derived from other kinds of waveforms. Why not record the video on the disc in digital form as well?

I noted earlier that the superb fidelity of the compact disc has strained the other components of the audio system, often prompting us to replace our amplifiers and loudspeakers with higher quality components that take advantage of the CD. The equivalent in the video world is a higher-quality television receiver that would take advantage of the improved video fidelity that digital storage can provide. But it hardly seems feasible to market a high-fidelity video display just for digital LaserDiscs. It makes economic sense to use the household TV for this purpose. Until there is an improvement in the quality of broadcast television and a corresponding improvement in the fidelity of the home receiver, much of the inherent quality of digital over analog optical recording is wasted. As long as our home TV sets themselves have limited quality, they are the limiting factor in the home presentation, and the big selling point of the optical disc must be its superior sound.

The solution is called *high definition TV*, abbreviated *HDTV*.

High Definition TV

There is mounting pressure to improve the quality of broadcast television with HDTV to achieve quality comparable to that of the optical motion picture. This means increasing the number of lines in the TV scan from the current 525 to more than 1000 and increasing the resolution along a scan by a similar factor of two, thus giving a factor of four improvement in overall definition. This means that an HDTV signal on a 38-inch screen would have the same quality as today's TV on a 19-inch screen.

What is keeping HDTV from becoming a practical reality? One problem is bandwidth. With twice the number of lines in the picture, the bandwidth requirements are doubled. There is always the possibility of compressing the signal to reduce the amount of bandwidth needed. However, the likelihood of being able to compress the signal enough to fit into the currently assigned 6- Mhz TV channels, while retaining full-motion capability, is very low. Since the FCC is unlikely to assign more bandwidth for TV broadcasting, the ultimate solution is not clear. Because of the bandwidth problem, it is likely that the first distribution of HDTV signals will be via cable, since cable is now fairly widespread in major population centers and has the additional capacity.

Another broadcasting problem is compatibility. When FM stereo broadcasting was introduced, the FCC insisted that the technique used permit people with monaural receivers to receive the stereo broadcasts. The same thing was true when color television broadcasting first appeared. Again the FCC required that the scheme used be compatible with black-and-white receivers. A person with a black-and-white TV had to be able to receive a color broadcast in black and white. The FCC has set similar compatibility requirements for HDTV: the broadcast HDTV signal must be receivable on a standard TV receiver with the definition of current TV broadcasting.

Still another issue is the cost of HDTV receivers. The principal challenge is the manufacture of large-screen displays that can take advantage of the increased definition and thus convert your living room into a movie theater.

Because of all these technical and economic issues, neither United States nor international standards for HDTV have been agreed upon. But no matter what the standard, HDTV, as it is currently understood, will be analog.

One of the problems in arriving at a standard for HDTV is finding enough bandwidth to permit its broadcast. The problem would be aggravated for digital HDTV. It is difficult to send digital voice in the bandwidth of an ordinary voice channel. It's just the same with video. Nevertheless, digital video broadcasting would have many advantages. In addition to picture quality, there would be the ability to store the broadcast picture and then process it later in various ways. For example, you could zoom in on selected portions of the picture and enhance the image to bring out certain desired features. Your television receiver would become a computer—or, perhaps more accurately, your home computer would have the ability to receive, process, and display television pictures.

If bandwidth is such a problem, how would digital TV be distributed? Probably through optical cable, which has far more capacity than the electrical cable now used for standard TV distribution. But there are major economic considerations involved in a massive wiring effort.

The potential advantages of digital TV are so intriguing that some people are proposing that the whole analog HDTV process be bypassed in favor of an all-out effort to develop digital TV. Since this might take 10 to 20 years, stopgap improvements to existing television quality would be made to satisfy the demand for higher quality in the interim.

Despite this school of thought and despite the lack of standards, analog HDTV is probably coming, first in Japan and Europe and later in the United States. Digital TV will come too, but it is very hard to predict how long it will take.

The Interactive Compact Disc

We will see recorded digital video via compact disc technology long before digital TV broadcasting becomes practical. The next household word in recording technology may well be *CD-I,* which stands for *compact disc interactive*. The CD-I looks like a CD; it is the same size and color. The CD stores digital audio only in one particular very-high-fidelity format, and a CD player's one function is to extract the stored audio digits and convert them into audio signals. The CD-I, on the other hand, can store digits in multiple formats to represent audio at different quality levels and to store video at different quality levels. What's more, the CD-I can store these different formats on different parts of the same disc. The consumer interacts with the disc by selecting that portion of the disc that he or she wants to play. The consumers will be able to use their existing video and audio equipment to display the video and play the audio. Thus, the video quality levels that can be accommodated by the CD-I are compatible with the current TV standards, as well as with the higher resolution standards that will be available with HDTV.

One of the key ingredients of this new technology is flexibility. With a flexible new technology it is not always easy to predict where the technology will ultimately find its greatest application. It has great educational possibilities that could take advantage of the flexible voice and video storage possibilities. Similarly, such a device could be very useful as a computer peripheral device. It will be fascinating to observe the applications as they unfold.

Computer Storage on Compact Discs

If a medium can be used to store audio and video, the same medium can be used to store digits that represent arbitrary data, which is, of course, the function of computer memories. The compact disc is finding application as a computer read-only memory, better known by its abbreviation, *CD-ROM,* because a huge amount of data can be stored in a very small space. A good example is the 20-volume Electronic Encyclopedia stored on a standard 12 cm-diameter disc. In another more recent example, the entire 26-volume edition of the Encyclopedia Britannica has been stored on a single CD-ROM complete with sight, sound, and new indexes that could make a significant impact upon library research in the years to come. The CD-ROM can hold as much data as around 1500 of the standard floppy disks customarily used with personal computers. The CD-ROM differs from the entertainment CD only in that the error-correction parameters are changed to provide still greater accuracy, which is not needed for audio storage.

For encyclopedic material, a read-only storage medium is preferable to a read-write medium because it is immune to accidental erasure. However most memory applications do require the ability to store as well as retrieve. A novel form of optical storage called the *erasable compact disc* provides this capability. We will discuss erasable CDs shortly.

Digital Audio Tape

Magnetic storage and magnetic audio recording appear in many contexts. The magnetic tape recorder stores replicas of acoustic signals by producing a track on the tape material that is magnetized in proportion to the strength of the acoustic signal. You listen to the tape by running the track across a magnetic head that senses the intensity of the magnetization along the track and generates a proportional electric signal. Magnetic tape has long been an alternative to the vinyl record, with the added ability to record signals as well as play them back. In contrast, once a vinyl record disk is cut, it remains a permanent record of the audio material. It cannot be erased and used to copy another audio signal.

Magnetic tape is widely used to store digital signals in a similar way. Instead of allowing the intensity by which the tape is magnetized to vary continuously to create an analog representation, a small spot on the tape is magnetized at one of two levels corresponding to a binary 0 or 1. A magnetic tape stores a sequence of bits in these magnetic spots in much the same way a compact disc stores a sequence of bits in pits on the disc's surface.

Digital magnetic tape has been with us since the 1950s. Bits can be stored and retrieved very rapidly on magnetic tape. But since it is a sequential-access medium, it can take a long time to retrieve a collection of bits at one end of the tape if the other end happens to be positioned under the playback magnetic head. It takes less time, on the average, to retrieve data from a magnetic disc than from a tape. And because of this, disks are the most common secondary storage medium, with tapes used as a back-up storage medium for archival purposes.

An inherently sequential-access medium is uniquely suited to audio recording, and both analog and digital audio tapes are widely used. Digital tape has long been the medium of choice for the initial recording of audio material. Tapes and records are copies of the digitally recorded master tape. But the tape recorders used for both computer storage and studio audio recording are large and expensive. For home entertainment purposes, a small cassette device that could be mass-produced at prices comparable to those of analog cassette recorders or CD players is desired. Such a device could be used to record compact discs, much as an analog cassette tape recorder is used to record LPs.

A digital recording made from a radio broadcast would be no better than a high-quality analog recording, since the inherent broadcast quality is generally poorer than that of even analog recording devices. But there is no reason that a digital magnetic tape recorder could not make very high-quality copies of CDs. The digital magnetic tape as a read-write storage medium has the potential to be a very valuable part of a home entertainment center.

The *digital audio tape* medium, with *DAT* as the inevitable acronym, was developed for this application. It uses credit-card-size cassettes in a small machine. However, a very long time elapsed between the development of these devices and their introduction into the U.S. market. The reason for this has nothing to do with their quality or cost, but rather with the fact that the recording industry, fearful of wholesale CD copying that could potentially impact CD sales, was successful in keeping them off the market until an agreement was reached on the capabilities of the machines.

The industry had similar reservations about video tape. It would kill the movies, they said. But not only are movies as strong and profitable as ever, but business is booming in movie cassettes designed for the home market. It is very difficult to predict the effect of new technology on the marketplace.

The recording industry initially attempted to insert digital codes on the CDs that, when detected by a special circuit on the tape recorder, would prevent the discs from being copied onto the digital tape. All these attempts were unsuccessful. Even had someone developed such a technique, it would not have been long before some resourceful young engineers would have developed and sold, legally or not, antidotes to such circuits. Artificially withholding new technology may work for a while, but inevitably it will be defeated.

The standardized format originally developed for recording digits on audio tape is different from the format used on compact discs. Most significantly, the audio sampling rate is slightly different, which means that this format does not permit the audio digits retrieved from the CD to be recorded on the tape. Rather, it requires the digits from the CD to be converted to analog and then redigitized before recording. This would degrade the quality of the tape-recorded audio, but not noticeably to all but the keenest audiophile ears. However, the chain of identical digits that marks digital communication would get broken in the process. If, for some reason, the digital signals on the tape were converted back to analog and then redigitized, there would be another slight degradation in quality. In general, the more such conversions, the more the quality degrades.

The agreement reached between the tape manufacturers and the record companies in the summer of 1989 to allow the DAT machines to be marketed in the United States stipulates that the machines will be able to copy CDs, but will not be able to make copies of digital tapes derived from a CD or other digital source. Thus, you will be able to make as many single digital tape copies of a CD as desired, but you will not be able to mass-produce any one of those tapes. Most significantly, the standard machine agreed upon by both industries provides a direct-digital mode at the CD sampling rate that copies the CD's audio bit stream without the need to convert to analog.

Rerecording on Compact Discs

The compact disc may yet give audio tape a run for its money as a versatile recording medium, either for entertainment purposes or for computer storage. Techniques have been developed that allow the data recorded on discs to be erased, thus converting the CD from a read-only memory to a read-write memory. To do this, the storage mechanism must be different from the standard mechanical process in which pits are etched permanently into the material. The read-write process, like the read-only process, uses a finely focused laser beam, but there the similarity ends. The read-write disc contains a magnetic layer in which the direction of magnetization is modified by the heat generated by the laser beam. Some materials are very difficult to magnetize at room temperature, but become much easier to magnetize when the temperature is raised. Therefore, when a low magnetic field is applied at room temperature, the magnetization of the material is unaffected *except* in the very small spot illuminated by the laser in which the temperature has been elevated. In that spot, the magnetization is reversed—a 0 becomes a 1 or a 1 becomes a 0, depending upon the direction of the magnetic field.

The digits are also retrieved from the disc by a laser, this time at a power level so low that the temperature of the illuminated spot is unchanged. The light beam distinguishes between a 0 and a 1 by polarization. Any electromagnetic wave, whether radio or light, has an electrical field that vibrates at the frequency of the wave and in a direction determined by the source of the original vibrations that generated the wave. For example, in the case of radio waves, the direction of the vibration is determined by the direction of the transmitting antenna. When you adjust the rabbit-ears antenna on your TV, you are attempting to adjust its polarization to match that of the received wave. Ordinary sunlight is the product of all the unrelated atomic vibrations in the sun and is therefore polarized in all directions. Polarized sunglasses reduce glare by admitting light with only a single direction of polarization, rejecting all other polarizations. Unlike the sun, the light emitted by the laser has a single direction of polarization. The magnetic material on the disc reflects this light, but shifts the direction of polarization according to the direction of magnetization at the spot from which the light is reflected. When the equivalent of polarizing sunglasses are inserted in front of the light detector, it can tell the direction of magnetization of the reflecting spot and, hence, whether a 0 or 1 was stored.

These erasable discs were first used commercially in a computer application. The pioneer was Steven Jobs, a founder of Apple Computers, who left the company in a well-publicized disagreement with the new management and started a new company called NeXT. His first product, a computer work station (the high end of the desktop computer market), uses erasable CDs in place of the usual hard magnetic disks because of the enormous storage capacities achievable in a small space. It may be a matter of time before they reach the entertainment market and compete with the new DAT technology. The economic issues that dominated the confrontation between the record producers and equipment manufacturers over DAT technology are exactly the same here. One would think that a similar agreement in this case could bring erasable discs to the entertainment market.

Future Trends

It is important to remember that the success of a new technology depends not only on the ingenuity of its inventors, but also on the public acceptance of the technology's applications. When a new technology is introduced, its price is high. If the technology becomes popular—as did the CD, for example—mass production causes the price to drop rapidly. If not, prices will fall slowly, if at all. There have been spectacular failures of potentially remarkable technology—the Picturephone is a good example of a technically successful technology that never found acceptance in the marketplace.

How eager is the public for a digital audio recording capability? The recording industry, in its attempts to keep DAT off the market, took the view that large numbers of people would run out to their neighborhood audio store, buy a DAT machine, copy all their CDs, give them away or sell them, and drastically cut into CD sales. As the DAT machines reach the market, they will be expensive, and they will become affordable to the average consumer only if they really catch on, as did the CD. Similarly while NeXT is using erasable CDs in its computers, it is not clear what impact the technology will have on the audio market.

Compact discs and digital tape are examples of storage technologies that have specific applications to both computer and entertainment systems. The CD-I is fundamentally different because it bridges the two areas. It may well be the first in a long line of such applications of technology and, thus, potentially the most influential of all.

9

Public Telecommunications

The telecommunications market place has undergone radical change in the postwar decades. This change is the result of a complex combination of legal, entrepreneurial, economic, and technological issues. The dominant technological trend has been the steady conversion of the public telecommunications networks from analog to digital. There is no better example of the mutual dependence, and, perhaps, ultimate fusion of the disciplines of telecommunications and computing than in this digital evolution of the public networks.

The Changing Networks

It is hard to imagine that not too many years ago most Americans bought all their telephone service from one company, The American Telephone and Telegraph Company, known to all, with appropriate filial affection, as "Ma Bell." Like a proper mother, Ma Bell took care of all the needs of her children even down to the ubiquitous telephone instrument itself. It was considered a radical change when we were finally given the opportunity to buy our telephones from competing vendors. Now we have the choice of buying all kinds of sophisticated customer-premises equipment from a host of competitors or to continue leasing an increasing variety of services from local and long-distance telephone companies.

As a result of the steady trend from monopoly to competition, culminating in AT&T's divestiture of its local operating companies, the telecommunications price structure has also changed radically. In the days of the AT&T monopoly, the cost of providing a service and the price charged to the public were only loosely related; cross-subsidies were the rule. Today's subscribers are more likely to pay for what they get but are not always happy about it.

The immediate post-war AT&T provided analog services only, almost all of it voice, in an all-analog network. Now, many companies offer a variety of voice, video and data services from a set of networks that are evolving from analog to digital. Most of this digital evolution is not evident to the average telephone system user. While the carriers are quick to advertise their massive digitization efforts, the effects of this evolution are largely hidden, because the commodity that the subscriber buys from the telephone company is still an analog circuit. It is also difficult to recognize because the changes have been slow, the very definition of an evolution as opposed to a revolution. The competing long-distance companies advertise how their digital services improve quality, but slow improvements in quality are sometimes hard to appreciate. Similarly, the lower prices to the subscriber have been masked by the massive changes to the whole cost structure that have resulted from divestiture.

Since information is a commodity vital to the way we live our lives and conduct our business affairs, changes in telecommunications beget societal changes. As profound as the societal changes wrought by telecommunications changes have been, they are trivial when compared to what is likely to happen in the decades to come.

Circuits

It is only natural to think of a circuit as a physical connection, a piece of copper wire between a pair of subscribers. Often, this is indeed the case. When your local telephone company installs service in your home or business, it literally runs a pair of copper wires between its *end office* and your premises. Long-distance circuits are not quite so straightforward. They may be microwave radio circuits spanning hundreds or thousands of miles in a sequence of relays, satellite circuits spanning thousands of miles, or fiber-optic cables. Or they may be a combination of several such media. While each of these circuits has its advantages and disadvantages in different circumstances, from the subscriber's point of view they are equivalent: each provides a way to deliver signals from one place to another.

Switching: How a Telephone Call Happens

The public communications networks are complex. In addition to *transmission facilities,* the name given to the collections of circuits of all kinds, they also contain another essential set of elements called *switches*. For this reason, they are called *switched networks*.

Consider the problem faced by a telephone company. It has millions of subscribers, any one of whom may want to talk to any other at some time. The most straightforward way to do this would be to run a telephone circuit between every pair of telephones in the country. However, with more than 100 million telephone locations in the country, it would take at least 500 trillion (500 followed by 12 zeros) telephone lines to interconnect every pair, an impossible task. Even if it were feasible to run all those cables, it would be terribly inefficient and expensive. Any given subscriber will have occasion to talk to only a very small fraction of the other subscribers. Therefore, most of those lines would never be used, and even those that were used would be idle most of the time.

The solution to this massive problem is a *switched system* that permits the resources of the network to be shared among all the subscribers, rather than being dedicated to the sole use of individual subscribers. Switching has been an integral part of telecommunications from the earliest days. The original switch was a human at a manual plugboard. The subscriber would tell the operator to whom he or she wished to talk, and the operator would provide the physical connection. Switching today is automatic, and Figure 9-1 shows its basic principles. In Figure 9-1a, eight subscribers are connected to the switch by phone lines. When a subscriber dials (or punches in) a number, the switch first interprets the number and then makes the connection between the lines of the two subscribers. This is a simplified version of the way your local telephone company's end office works. Suppose that your telephone number (without the area code) is 456-3377. The *exchange* part of your number (456) designates a telephone company end-office facility in your neighborhood. Your telephone and those of all other subscribers in the 456 exchange are connected directly into the end office. When you call someone with the same exchange, the end-office switch makes the connection between the two phones, as shown in Figure 9-1a.

The advantage of switching is clear, if we consider a few num-

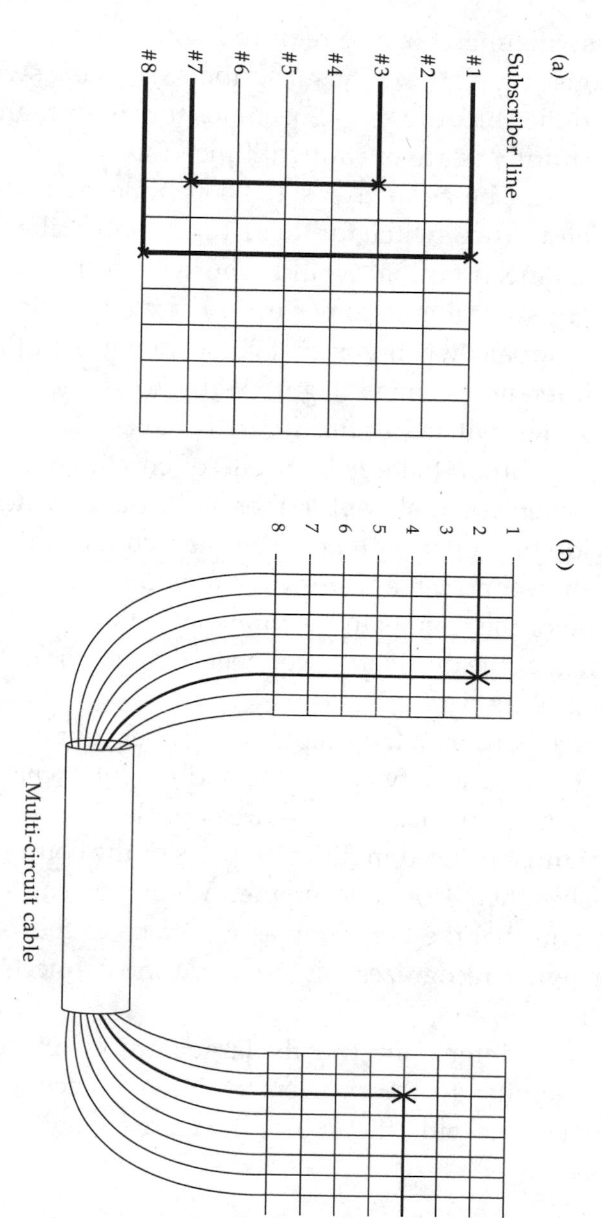

Figure 9-1 Telephone Switch Connections

bers. Without switching, it would require 28 lines to provide all possible connections among the eight subscribers. With the switch, only 8 lines are needed. The differential becomes much more impressive when realistic numbers are examined. Since four digits follow the exchange, there can be as many as 10,000 phone numbers at that exchange. An end-office switch for these 10,000 subscribers replaces the 50 million connections that would otherwise be needed.*

The power of switching becomes even more dramatic when you consider what happens when you dial someone on the other side of town with a different exchange. Figure 9-1b shows the principle involved. Each of the switches in this figure behaves like the switch in Figure 9-1a. In addition, there is a bundle of circuits connecting the two switches. When you dial a subscriber of the other switch, the two switches provide the connection between the two telephones using a circuit in the connecting cable. The switch in your end office connects your line to one of the circuits in the interconnecting bundle, and the end-office switch at the destination connects that circuit to the person whom you are calling.

The exchange portion of the telephone number contains only three digits allowing you to specify at most 1000 different exchanges. Since there are many more than 1000 local areas in the United States, the same numbers must be used in different parts of the country. The *area code* distinguishes them from one another. When you call someone in another area, you dial the area code as a preface to the 7-digit local number. Your switch recognizes this area code and connects you to the correct end office.

The problem of interconnecting the large number of end offices in the country is similar to the problem of interconnecting the many subscribers of a single end office. Since there are so many end offices,

* The formula for computing the number of pairs of N objects is $1/2 \times N \times (N-1)$. When N is 8, the result is $1/2 \times 8 \times 7$, or 28. When N is 10,000, the result is $1/2 \times 10{,}000 \times 9999$, or 49,995,000.

it is just as impractical to provide direct connections among all pairs. Additional switching is used to provide this interconnection. Each long-distance company has built a hierarchy of switches throughout the country designed to perform this complex interconnection efficiently.

The principle behind switched service is resource sharing. Rather than dedicate facilities to individual subscribers who may use them only a small fraction of the time, it is more economical to build a network that shares facilities among many subscribers and thereby minimizes idle time. However, a price is to be paid for this efficiency in terms of network availability. The ordinary subscriber isn't usually aware of this limitation, but might notice it at a time of unusually heavy usage such as Mother's Day, when it might take several attempts before a call goes through. The switched network has some maximum call-carrying capability, based upon the number of calls that each switch can handle at once and the number of circuits interconnecting the switches. If the number of calls attempted exceeds this capacity, some of the attempts will not be successful.

The telephone system has enough capacity to handle normal daytime business traffic volume, but is underused on nights and weekends. This is why the carriers encourage the use of their otherwise idle plants by offering lower long-distance rates at those times. But on the relatively rare occasions of unusually heavy demand, such as Mother's Day and during floods and earthquakes, the traffic may exceed the normal business day traffic, resulting in a deterioration of service. If the networks were sized for these atypical circumstances, they would be underused most of the year, and the rates would reflect that. It's much like a city expressway designed to cope with normal rush-hour traffic becoming immobilized by the occasional emergency that creates peak traffic demands. While we are all prone to grumble when these occur, few of us would be willing to pay the price of a highway system that could cope with such emergencies gracefully.

Analog and Digital Circuits

A Telephone Company View

What appears to the subscriber to be a very simple telephone connection is, in reality, a very complex network of circuits and switch connections. This complexity has much to do with the quality of this long circuit, because every complexity in the communications path is an opportunity for quality degradation.

As shown in Figure 9-2, these circuits can be analog, digital, or a hybrid of the two. An analog circuit is analog from end to end; all its constituent links are analog. Similarly, a digital circuit is digital from end to end. In a hybrid circuit, some of the links are analog and some are digital. Today, by far the majority of the links connecting subscribers to their end offices are analog, because they are there primarily to serve analog telephones. Since much of the rest of the network is digital, the hybrid circuit today is the rule rather than the exception.

What can we say about the accuracy or quality of such links? The situation with pure analog and digital links is clear. An all-analog circuit suffers more degradation the more links there are in the chain. In contrast, the all-digital circuit has high accuracy no matter how many circuits are linked. A hybrid circuit is somewhere in between. In the most typical case, shown in Figure 9-2, where the subscriber "tail" circuits are analog and the rest are digital, the circuit quality is ultimately determined by the quality of the tail circuits. Wherever the characteristics of these are well controlled, the circuit quality is high; wherever they are not, the quality suffers. Since most of the circuit degradation is contributed by the local analog connections, a long-distance circuit is not very different in quality from a local circuit. Many express surprise when a voice coming from thousands of miles away sounds as if it were "next door." The reason is the high-quality digital connection between end offices.

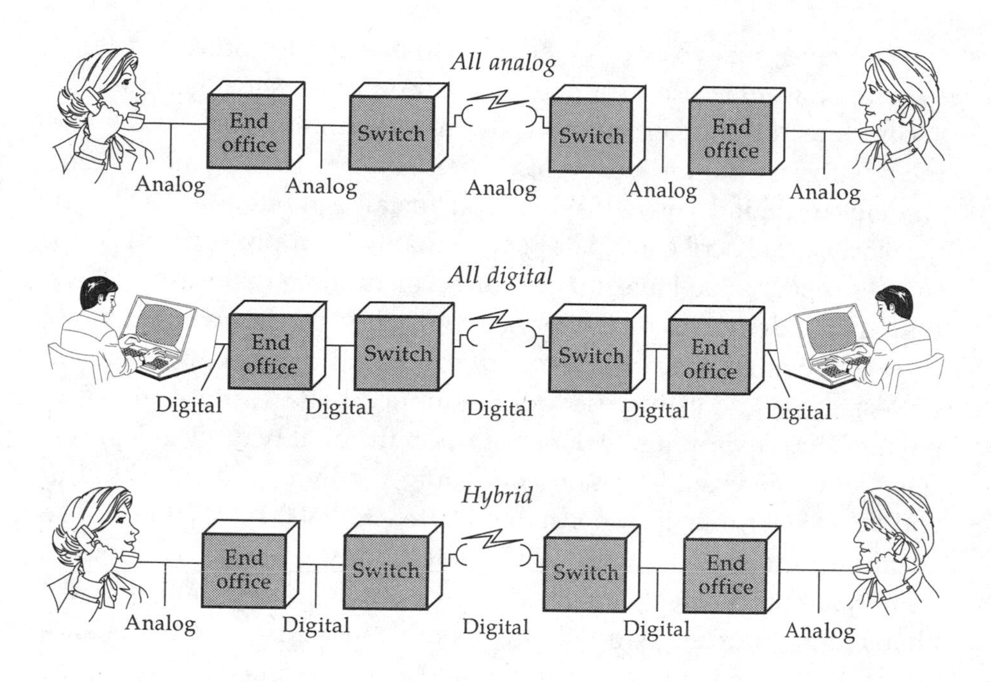

Figure 9-2 Circuits in the Public Switched Networks

A User View

The question of what is analog and what is digital in a telephone system depends upon who is looking, the carrier or the subscriber. We have just observed circuits from the point of view of the carrier. When we look at telecommunications from the user's point of view, the perspective is somewhat different.

Local telephone company subscribers who are connected to analog circuits are, in effect, buying bandwidth. Most subscribers want voice communication and nothing more. For that purpose, the telecommunications industry has defined the standard telephone *voice-frequency channel* with a bandwidth of about 3 kHz. It was originally designed for voice service and has continued as the dominant telecommunications commodity.

However, this voice channel can also be used for other purposes. When you want to use your personal computer to exchange data with a computer in another location, the most common vehicle for this interchange is the ordinary voice channel. You convert this analog channel to digital yourself when you install a modem at each end. Similarly, to subscribe to a data service that will supply you with stock market reports, you have only to connect a modem that is compatible with the modem used by the service company, and the analog channel provided by the carrier is converted to a digital channel. But there are data-rate constraints imposed by the limited bandwidth of the voice circuit. The maximum rate depends upon the quality of the telephone line and the sophistication of your modem and its built-in coder/decoder, but it is normally limited to 2400 or 4800 bps. The quality of the circuit varies. When you simply dial up a connection using your normal telephone service, the quality is determined by the characteristics of the two tail circuits and the variable long-distance circuit that happens to be given to you by the switched network. The more of the system that is digitized, the better will be the average quality of your service.

While switching provides the enormous economic advantage of resource sharing, there are times when it is economical for a subscriber to pay the additional price of a dedicated circuit to one particular location. A *dedicated circuit* is guaranteed to be yours 24 hours a day, seven days a week. For your application, it may be worth the cost to have the circuit all to yourself, eliminating the risk of a busy circuit. Sometimes, it is even more important that a dedicated circuit offers the ability to select the line quality. Higher quality lines with improved bandwidth characteristics and higher signal-to-noise ratios can yield higher data rates, perhaps as high as 14,000 bps for a commensurate price.

Whether the service just described is switched or dedicated, it remains analog and is still constrained by the 3-kHz bandwidth. But sometimes you can buy end-to-end *digital service* directly from a carrier or combination of carriers. In this case, you are buying not bandwidth per se, but rather a certain data rate with a guaranteed maximum error rate. The telephone company supplies all the modem equipments necessary to bring digital service to your door. You needn't worry about choosing the best modem for your application; that's all done for you by the carrier. This kind of service is not very common today, but is expected to expand rapidly as time goes on.

Switched Data Service

When the networks were all analog, it was common to use the terms *digital* and *data* synonymously. Now we must be more precise, since data can be transported by either digital or analog networks. A *digital service* is one that, in contrast to an analog service, allows you to send your data at a variety of rates without the need for modems external to the network. A *data service* is a service designed to support data transmission, usually computer-to-computer or person-to-computer, rather than voice transmission. While the carriers usually implement a data service with digital transmission internal to the network, an individual subscriber usually gains access to the service through an analog circuit, perhaps dial-up, requiring an external modem.

Like voice service, data service may also be switched. The advantages and disadvantages are the same as for switched voice service: the economy of resource sharing is an advantage, but when demand exceeds supply, service is degraded.

There are some inherent differences in the nature of the switched services needed for voice and computer data. Our telephone switching systems are optimized for voice communications. It takes a few seconds to make the connection after dialing the number. Once the connection is made, you use the circuit continuously until the conversation is finished, however long that might be. Only then do you relinquish the circuit. The carriers typically design their network capacities using four minutes as the average duration of a business call. This type of system is called *circuit-switched.*

On the other hand, if you are at a computer terminal and want to exchange data with a computer interactively, you dial up the computer, and your terminal and the computer alternately send data to each other. The session might last for hours rather than minutes, but within that session, data might be transmitted only sporadically, a few seconds at a time, with large gaps in between. For example, you might send a brief message to the computer. The computer might respond quickly with a burst of data sufficient to fill your monitor screen. You might not respond for minutes or longer. This type of transaction calls for a system that will provide rapid communication when either the terminal or the computer have data to send. At all other times, the resources of the network might just as well be used by someone else, provided that your connection can be restored on a moment's notice.

A circuit-switched system could do this if it were able to set up and take down a circuit in a fraction of a second, far more rapidly than is now possible. Otherwise, the time to restore the connection could be a significant fraction of the entire transaction time, a situation that a subscriber would deem unacceptable. For these reasons, *packet switching* has been devised with characteristics optimized for interactive data transfer. The data in a transaction are divided into short segments called packets. Instead of dedicating a low-rate circuit to each pair of users, the packets from all users on a given route share a common high-data-rate path. These packets are sent over the path at very high speed in the order in which they are received at the switch. Packets may have to wait their turn by an amount of time that depends upon the instantaneous traffic load, but that is no problem unless there are unusually heavy traffic loads. This system gives the user an illusion of full-time connectivity, yet reserves network capacity only for the time taken to actually transmit the data.

There are many packet-switched networks in operation today. The father of them all, called the *ARPANET*, was built by the Defense Advanced Projects Research Agency (DARPA) in the late 1960s and was in operation until early 1990. But because the total volume of data traffic is still so much less than the volume of voice traffic, the data networks are still much less significant than the voice networks in the total communications picture.

Packet-switched service is primarily for interactive data transmission and similar applications. A circuit-switched system will work just as well or even better for massive bulk data transfers between computers. For digital voice, the time that each packet must wait at the switch can present a fundamental problem, since each packet is delayed a slightly different amount of time. If uncorrected, this would have a disastrous effect on speech quality. Therefore, to make packet switching work for voice, these differential delays from one packet to the next must be compensated for, so that they are not evident to listeners. While this has been done experimentally, it has not been applied by the public networks.

It appears that circuit switching is the more natural way to handle the characteristics of voice communication, either analog or digital, and packet switching is more suitable for interactive data transmission. However, if a single switching system could handle all types of traffic, there would be economic advantages. Should such a system be circuit switched, packet switched or even some hybrid combination? As it turns out, there are passionate advocates of circuit switching for everything, and there are equally passionate advocates of packet switching for everything. The last word on this topic has not yet been written.

Private Networks

Aggregation and resource sharing mean economy. For this reason, the most economical approach for the subscriber with modest communications needs is to take advantage of the economies of scale achieved by the switched service offerings of the public networks. But sometimes, an individual company or a government agency may have such extensive requirements that a switched network of its own is the most economical approach. Such a *private network* would serve the company's own locations, but would also interconnect with the public networks to allow access to outside locations. A private network is similar to a public network, with various kinds of switches and transmission facilities, but all dedicated to a particular user community. Some government examples are AUTOVON, the Defense Department's AUTOmatic VOice Network, and the Federal Telecommunications System (the most recent version is called FTS-2000) administered by the General Services Administration for the civil side of government. Large corporations such as IBM, General Motors, and Ford have had their own private networks for some time.

A private network may be procured in many ways. A company may lease the network from a carrier or may buy the equipment and operate it itself. The company may buy its own switches and lease the transmission facilities from one or more carriers. For example, many companies of all sizes have found it economical to buy on-premises switches to serve as local switchboards for internal telephone service. These same switches, called *private automatic branch exchanges* (abbreviated *PABX*), can be used in a slightly enhanced form to interconnect locations, thus constituting a complete intracompany network.

The public networks have responded to the challenge of this competition by taking advantage of their digital networks and smart computer-based switches (to be described later) to offer *virtual private networks.* Such a network is carved out of the public network, but in such a way that it looks like a private network to the customer. For example, AT&T calls its virtual private network a "software-defined network," emphasizing the fact that its capabilities are obtained by their ability to modify their switches through software changes. This approach can often be more economical than public-network service for moderate-sized companies with requirements too modest to warrant a totally private network.

The Digital Network Evolution

The public networks constitute, in the main, large circuit-switched systems that were initially completely analog and have been evolving toward digital internally, while maintaining analog interfaces to the user community. There are many reasons for this internal digitization. One is the controlled quality. Another is the steadily increasing user demand for data transmission at high rates not easily obtained from the uncertain quality of an analog system. Other, less obvious factors related to flexibility are becoming increasingly important. Users have been attracted to what are called *enhanced services* associated with both voice and data service: call waiting, call forwarding, conference calling, and other similar convenience features. It takes much less time to make a long-distance connection than it used to, and this, too, is a direct consequence of digitization. The ability to carve virtual private networks out of the public network is still another advantage.

But the most important reason for digitization is cost. Isn't this like putting the proverbial cart before the horse? Usually a new technology is introduced to obtain new services, with the expectation that the revenues derived from the services will exceed the cost. The remarkable thing in this case is that the economic benefits of digitization would be of great significance in their own right without the abovementioned service benefits.

Digital Switching

Today virtually all switching, both circuit and packet, is digital. The first automatic switches emulated manual switches by interpreting the dialing information and then making physical connections between circuits using relays. With the advent of digital computing technology, these old switches were gradually replaced by more modern versions that took advantage of the flexibility and economy of the computer.

The modern communications switch, whether it be a circuit or a packet switch, is, in essence, a digital computer with certain specialized capabilities to improve its ability to do switching.

An analog switch works by making a physical connection between two circuits. A digital circuit switch works differently. Each incoming circuit is treated by the digital switch as a computer treats an in/out device such as a keyboard or a display terminal. The digital switch accepts data from one circuit and then, after some manipulation, delivers the data to the other. The switch must do this quickly enough for the connection to be indistinguishable from a physical connection.

The public network switches were, for the most part, converted to digital while much of the transmission was still analog. At first glance that seems peculiar. For it means that the analog signals from the transmission must be converted to digital before they enter the switch. Similarly, the digital signals leaving the switch must be converted back to analog to be compatible with the analog transmission. Even with all this signal conversion, the carriers chose to build their systems in this way because of the economy and flexibility of the digital switches.

This revolution in communications switching was the direct consequence of the rapid strides taken by the digital computer. The same technological forces that permit you to afford to buy a home computer today with the power of a large mainframe of ten years ago permit the phone companies to greatly reduce the cost of their switching. Many of the benefits derived from the use of flexible digital computers are therefore obtainable as a consequence of the switches going digital rather than as the driving force behind digitization.

The costs associated with switching come in many forms. Most obvious and the simplest to understand is the direct cost, i.e., the number of dollars and cents needed to buy the item in question. Another cost is maintenance: the all-electronic digital switches require less maintenance than the older electromechanical analog switches.

Another cost is physical size. As an example, AT&T chose to digitize the largest switches in its network first, beginning in the mid-1970s. The smaller size of the digital switches led to a significant reduction in the amount of building space required as the network expanded. So important was floor space in the cost equation, that the company found it economical to replace analog switches that had been in operation for only a few years, a very short time compared to the expected lifetime of a switch.

Digital Transmission

Digital switching improves economy, responsiveness, and flexibility, but not circuit quality. A network consisting of digital switches and analog transmission is cheaper than an all-analog network, and it offers some enhanced services, but its circuit quality is still that of an analog network. Only when the transmission is also digital do we gain the true quality benefits of digital technology. We also gain more switching economies, because all the conversion equipment disappears.

The evolution of transmission within the public networks is characterized by a quest for more bandwidth to satisfy an ever-increasing demand for capacity. As the capacity requirements of a network grow, its interswitch trunks must expand to carry more circuits, thus necessitating more and more bandwidth. To accomplish this, it is more economical to *multiplex* a number of individual circuits together into a single wideband circuit than to organize the bandwidth as a set of individual narrowband circuits. For example, a typical analog group carries 24 voice circuits. Similarly, a group of 24 voice circuits is handled as a single digital entity using a data rate of 1.544 Mbps, referred to as *T1 service*. When bandwidth is at a premium, the carriers may use one or more compression schemes to increase the circuit- handling capability of the existing bandwidth.

With analog transmission, the reason for grouping is largely economic. Less bandwidth is required when all the circuits in a group are handled as an entity, and a single wideband modulator or demodulator is cheaper than a set of individuals. The same economies are achieved with the digital grouping. But there is another issue as well. A high-rate digital channel can apportion its capacity far more flexibly than can an analog channel. In an analog network, the commodity is usually the 3-kHz voice frequency channel and, less frequently, the wideband channel suitable for television. In the digital network, a 1.544-Mbps stream may be subdivided in many ways. It may be used to carry 24 or 44 voice channels, but it may also support near-full-motion video for conferencing in 772 kbps and 12 voice channels in the remainder. Very-high-rate trunks that support multiples of the T1 rate have virtually unlimited flexibility to carry wide varieties of services at different rates.

Whether digital or analog, interswitch transmission is a large consumer of bandwidth, and the bandwidth itself may be the determining factor as to whether digital or analog transmission is used. For the same transmission capacity, digital service offers improved quality, but often pays for it by requiring more bandwidth than do certain types of analog service optimized to minimize bandwidth use.

Bandwidth limitations are always a problem, but the severity of the problem depends upon the medium used for transmission. Transmission media fall into two general categories: wire (cable) and wireless (radio), and the problems are different in the two categories.

Wireless Transmission Media

Figure 4-5 showed how the electromagnetic spectrum is divided among many diverse applications. The process of allocating it among these uses is very complex and is subject to many pressures from the various interests involved. Certain bands have been allocated to various communications services such as radio and television broadcasting and fixed and mobile radio services. Microwave relay is included in this category. Microwave antennas are located as high as possible on towers or the roofs of buildings to maximize the line-of-sight range. Signals are transmitted from one antenna to another, supporting the transfer of information in proportion to the available bandwidth. To meet the demand for communications capacity, the more densely populated areas of the country have been covered with microwave links, and, because of the limited available spectrum, they all must share the same limited bandwidth allocation.

Bandwidth is even more limited because nearby microwave links can interfere with one another. While most of the energy radiated by a transmitting antenna is collected by the intended receiving antenna, a very small fraction of the energy radiates in other directions, and some of this energy can enter the antennas of neighboring links. Therefore, the total band allocated to microwave relay must be subdivided, so that neighboring links can use different frequency bands and thereby avoid interference. But this limits the amount of bandwidth available to any given link. Accordingly, analog in its more bandwidth-efficient forms continues to be widely used on many of these microwave links, despite the superior quality achievable on digital links.

Since satellite circuits may be thought of as very long microwave links, they are subject to these same limitations. In fact, some satellite links use the same frequency bands as do microwave links, and their applicability has been limited in urban areas because of mutual interference with microwave links. However, satellites have never been a major component of the transmission used by the domestic networks. They have been more widely used for transoceanic transmission where the alternatives are considerably more limited. In either case, their use is sometimes analog and sometimes digital, the choice depending upon a number of factors.

Cable Transmission Media

The situation is quite different with cable transmission. Whether the cables are electrical or optical, there is no interference problem. The only limitations are those inherent in the media themselves.

Originally, nearly all transmission, short- and long-range alike, was by electrical cable. The bandwidth limitations of the cables led to their ultimate replacement by microwave relay for all but the shortest links. Now, fiber-optic cable is the medium of choice for long-range circuits, because of the enormous bandwidth available at optical frequencies. All the long-distance carriers are installing optical cable, both to replace microwave on existing routes and to establish new routes.

A medium such as optical cable is especially cost effective on very high capacity routes. The single most important cost of cable is its installation, including the cost of the real estate for the cable path and the cost of laying the cable underground (the preferred method). These costs are the same, whether the cable carries a thousand circuits or a million circuits. Thus, it is most economical to design the network in such a way that maximum aggregation is possible. It seems evident that the circuit trunks between major population centers (e.g., New York to Chicago, etc.), as well as within these centers, will be very heavily aggregated. In recent years, the major local carriers have multiplied their capacities manyfold by replacing their electrical cable with optical cable within their existing rights-of-way. Similarly the long-distance carriers have installed very high density fiber-optic transmission trunks on major intercity routes to augment or replace existing microwave transmission. And with seemingly unlimited bandwidth at their disposal, the carriers have gone to digital universally in their application of optical cable.

Digits at Your Front Door?

Thus, the public networks have taken great strides in converting their networks to digital. Some of the largest users have been able to take advantage of door-to-door digital service at T1 rates and higher in their private networks. But most of this digitization is invisible to the average subscriber, who still has an analog interface with the networks. The limitations of analog phone service become evident only when a subscriber needs to transmit data at rates in excess of the capabilities of an analog circuit. If analog service is adequate for most subscribers, what are the compelling reasons to bring digital service to their doors?

One reason might be digital voice transmission; the 56- or 64-kbps standard does yield substantially higher quality than ordinary telephone-quality analog voice, and rates this high are not supportable with the existing analog tail circuits. But there is no compelling reason for this improved quality if its cost is high. The quality improvements possible with digital voice would entail a substantial, costly redesign of the subscriber distribution network, and it is not at all clear that the benefit merits the cost.

Another possible reason is security; the only way to make speech secure in the presence of a determined wiretapper is to digitize it and then randomize the digits in a predictable way. The technique is to encrypt the speech digits by adding them to a sequence of pseudorandom digits. Only the intended receiver knows these pseudorandom digits and so is the only person who can decrypt them. There are various kinds of scramblers on the market that will protect analog communications from a casual listener. But these can be penetrated easily by a determined adversary. While secure voice is essential in matters of national defense and in some sensitive commercial activities, the need is so sufficiently specialized that it is unlikely to motivate the carriers to bring digital service and its high data rates to your door to improve the speech quality now limited by the low data rates supportable by analog circuits.

If digitized voice is not sufficiently urgent to warrant end-to-end digital services, then perhaps the anticipated growth of high-data-rate requirements is. But this is a hard case to make. It would appear that the demand for high data rates is neither so great nor so universal as to force the widespread installation of digital services on the customer's premises. Those companies desiring it have installed their own private networks with T1 service or have been content to live with the low-rate services easily obtainable with analog service.

Although the carriers have many reasons to digitize their networks, none alone is strong enough to justify the widespread installation of door-to-door digital service for voice, data, or video. But here is a case where the whole appears to be greater than the sum of its parts. The carriers are preparing to evolve to the integrated services digital network. ISDN is the abbreviation for this network that will bring digital services direct to your door, services to be used for voice, facsimile, data, video, and anything else that might come along in the future. The key words are integrated and digital. With ISDN, you will no longer have to buy a mixture of analog and digital services for your various needs—one circuit for voice, another for facsimile, packet-switched service for data, and specialized satellite service for video teleconferencing. When you buy the appropriate class of integrated digital service from your local carrier, you can have all these services with a single digital connection. For example, one of the standard offerings under ISDN will be 144-kbps service. This will give a user two circuits at 64 kbps per circuit, each of which can be used for either voice or data, plus 16 kbps for low-speed data. The network will then provide switched or dedicated connections using whatever technology is appropriate. It might, for example, send interactive data to a packet-switched network and voice to a circuit-switched network. How the carriers deliver the service is not important. What is important to the customer is that the communications carriers guarantee to meet a standard set of specifications at an affordable price.

The carriers are now pilot-testing ISDN services in a small number of locations both in the United States and in Europe, most notably in France. Large-scale installation of ISDN services would represent a considerable investment on the part of the carriers, and ISDN will become widespread only if there is sufficient demand. The first users will surely be large businesses where the demand will be the greatest. After that will come small businesses and later residences. The optical cable that could distribute digital TV would be able to provide ISDN services as well.

According to the industry, ISDN services should become increasingly available during the 1990s and almost universal by early in the next century. How rapidly will this really happen? Unquestionably the demand will increase with time, but it is anyone's guess when it will be sufficiently high to warrant large-scale installation of ISDN services.

The question of the viability of end-to end digital services is not new. In the late 1970s and early 1980s, some of the cable TV companies promoted ISDN-like services with their standard copper cables. But the demand at that time was so insignificant that the companies withdrew their offerings.

Still earlier, the Defense Department wrestled with the problem in its attempt to improve secure voice services. In the mid 1970s, an upgraded service was planned that was to be delivered by an end-to-end digital network. The choice of a digital network was made to gain the superior speech quality offered by the higher digitization rates, as well as compatibility with Army field telephones. The domestic portion of this service was to have been provided to the government by AT&T. At that time, the public network was still primarily analog, and to install the digital network would have necessitated a substantial refurbishment. The alternative was a lower cost vocoder approach using the existing analog plant. The debate that ensued between competing elements of the Defense Department and Congress was long and sometimes painful to those of us who participated. In retrospect, the most interesting philosophical issue was the extent to which the government could afford to lead rather than follow private sector developments. Ultimately the "followers" won the day. The digital approach was simply too far ahead of its time.

Public Telephone Service

The evolution to digital technology within the public telephone system has taken place side by side with a revolution in the way our national telecommunications systems are owned, are managed, and compete for our communications dollar. The stories of digitization and industry restructuring have been so intertwined that any discussion of one without the other leaves the overall story incomplete.

We are by now used to dealing with at least two companies for our telephone services, one local and one long-distance, whereas not too many years ago, we dealt with a single company. We are also used to the fact that several companies are competing vigorously for our long-distance dollars. How all this came about is a fascinating story; and two highly significant trends are behind it.

The first trend underlies the digital evolution; it relates to the symbiosis between computing and communications and, perhaps, their eventual union. Since a communications company also becomes a computer company as it digitizes its communications plant, why can't it also use these computing capabilities in other business areas?

The second of these trends is entrepreneurial. Was it necessary that the nation's communications services be provided by a near monopoly? Why was it not possible for other companies to provide similar services?

The tensions created by these trends over a period of several decades were largely unknown to the general public. Then in 1982, the courts ordered AT&T to divest itself of its local operating companies. This and the events that followed have affected every American and have brought the ins and outs of the communications business to the public eye as never before.

In most of the world the telephone networks are government-owned or chartered, typically with the establishment of a *postal telephone and telegraph company*, abbreviated *PTT*, on much the same basis as the United States Postal Service. This PTT operates all these information utilities on an exclusive basis. In the United States, while the mail service was chartered as a government-operated monopoly from the founding of the republic, first telegraph and then telephone service emerged in the private domain. Since a single company, AT&T, dominated the vast majority of this industry, the government regulated it to protect the interests of the consuming public, in much the same way as it regulates electric and gas companies that provide service to the public on an exclusive basis. Initially the Interstate Commerce Commission provided this function, but in the Communications Act of 1934, the Congress mandated the establishment of the Federal Communications Commission to oversee this regulation. The FCC's scope includes, in addition to price regulation of communications services, the domestic allocation of the frequency spectrum among competing services and users, a function regulated worldwide by the International Telecommunications Union. Since communications matters internal to individual states are not subject to federal regulation, each state has established a public utilities commission for this purpose.

From the beginning, however, there were small telephone companies independent of AT&T that provided local service. Early in the century the Bell System (AT&T's familiar name that originated from its control of the original Alexander Graham Bell patents) made a calculated attempt to obtain a complete monopoly over all telephone service by buying up many of these companies. In an antitrust action in 1913, the Justice Department succeeded in limiting this attempt severely. With AT&T out of the picture, many of the remaining independent companies merged over the years into larger companies such as General Telephone and Electronics (GTE), United Telecom, and Contel. But all these companies confined their services to restricted local geographical areas. Until the 1960s, AT&T was the only company providing long-distance services, in addition to providing the bulk of the local area services. Thus, even though GTE might have a local company in Virginia and another in California, AT&T provided the connections between subscribers of the two local GTE companies.

AT&T built an elaborate infrastructure. Not only did it operate most of the telephone service in the country, but it also manufactured most of the equipment needed for its network, from the switches, transmission and modems down to the telephone handsets, in a subsidiary called Western Electric Company. In its Bell Telephone Laboratories, AT&T maintained an extraordinary research and development program, extending from basic research in physics and mathematics to equipment development. It was here that Nyquist and Shannon did their pioneering work. It was also here that the transistor was invented, the device that formed the basis for the component revolution behind the whole information industry.

The rate structure philosophy established by AT&T, with the concurrence of the FCC, was an elaborate pattern of cross-subsidization in which long-distance service subsidized local service, urban service subsidized rural service, and the ratepayers, in aggregate, subsidized service in national emergencies. In the latter area, AT&T was able to respond very rapidly to emergency needs of the country without direct charge. The costs were simply incorporated into the rate base and ultimately paid for by all.

The Emergence of Competition

All this began to change in 1969, with the inception of a company called Microwave Communications Inc., now better known as MCI. With the approval of the FCC, MCI began to install microwave links of its own in the Midwest and to offer long-distance communications transmission services mostly to corporations at rates well below those charged by AT&T. A second change occurred when, the FCC encouraged industry to begin offering services to the public based upon the emerging communications satellite technology. A company could install satellite dish antennas on the customer's premises and provide private network services completely independent of AT&T and the local telephone company, AT&T or other. Several companies began to offer such services beginning in the early 1970s. Although all these competitive services came under the jurisdiction of the FCC, the start-up companies were free to charge whatever they liked. The rationale for this was clear. All the competitors together controlled an insignificant fraction of the total telecommunications volume in the country. It was, therefore, in their interest to offer services at lower rates than AT&T to gain some share of the market.

Another inroad into the near-monopoly exercised by the Bell system occurred in 1976, when the Southern Pacific Railroad began to build and sell long-distance service on a competing network, SPRINT. A railroad getting into the communications business is not as strange as it seems, when you consider that its railroad rights-of-way provided the real estate on which to construct the transmission capacity. SPRINT (later GTE SPRINT and still later one of the antecedents of US SPRINT) and MCI became bona fide competition to AT&T with networks that, by the early 1980s, began offering switched long-distance services directly to the public in many parts of the country. Even companies that, unlike MCI and SPRINT, did not own any transmission facilities of their own began getting into the act. Taking advantage of the great difference between AT&T's wholesale and retail rates, they would lease circuits from AT&T in bulk and resell these circuits to the public on an individual basis. This pricing structure also permitted such a *reseller* to buy switches and sell switched service to the public using the leased bulk transmission to interconnect the switches. Thus began widespread competition at the subscriber level.

The local telephone companies, whether owned by AT&T or not, were obliged to provide connections between their subscribers for these alternative long-distance carriers, even though these connections were generally of inferior quality. If you subscribed to one of those services you will recall, with some pain, that you had to dial some 20 numbers to make your long-distance connection. This was in contrast to the 10 or 11 digits required for normal AT&T long-distance service.

When this competition began, none of the alternative companies, especially the resellers, had networks that could approach AT&T's in extensiveness. Indeed, they deliberately confined their activities to high-volume geographical markets in which they offered lower cost services than AT&T, a process that came to be known as "cream-skimming." When the public demanded complete geographic coverage, the only way these companies could comply was by using facilities leased from AT&T in the low-volume areas. AT&T, being a regulated common carrier, was obliged to lease circuits to anyone—even to competitors. But by judicious application of "ratesmanship" AT&T was able to flex its giant muscles. For example, by lowering the price differential between wholesale and retail circuits, they eventually were able to do away with the resellers almost completely and force the larger facilities-based competitors to greatly expand their networks.

While this rapid expansion placed considerable financial pressure on the competitive companies, it had other effects as well. When these newer companies installed new switches and transmission facilities, they chose the most modern equipment available, which was almost always digital, buying their switches and other equipment from many competitive manufacturers both foreign and domestic. In contrast, AT&T was almost completely self-contained and self-sufficient and made all its own equipment. Both AT&T and the smaller companies were also quick to take advantage of the capacity and economy of the new digital fiber-optic transmission technology. But since the new transmission constituted a greater percentage of their networks than of AT&T's, the competitive networks have led the way in percent-wise digitization.

Divestiture

The Bell system was the subject of Justice Department scrutiny almost continuously for violations of the Clayton Anti-Trust Act. After all, despite the embryonic competition that we have just recounted, AT&T controlled almost all the public communications in the country, both local and long-distance, manufactured all its own equipment, and conducted an extraordinary research and development program. It is not surprising that AT&T recognized very early the impact that computing would have not only on communications networks, but also on society as a whole. The Justice Department was concerned that AT&T, with its near monopoly in communications, could convert that advantage to domination of the fledgling computer industry. In 1956, in what came to be known as the "Final Judgment," the courts reaffirmed AT&T's monopolistic position in providing *regulated* services only and, in return, required AT&T to get out of any other businesses, including the emerging data-processing activities.

The Final Judgment was well ahead of the emerging competition in long-distance services. A later antitrust proceeding was motivated both by the emerging competition in the long-distance area and by the increasingly incestuous relationship between the regulated telecommunications industry and the unregulated data-processing industry. The Justice Department proceedings began in the 1970s and finally culminated in the divestiture decision in 1982 that led to the break-up of AT&T. Under what is called the "Modified Final Judgment," the federal courts ruled that AT&T's dual position as the dominant carrier in both the local and long-distance areas inhibited the emerging long-distance competition. AT&T was forced to get out of the local telephone business by divesting itself of all its local Bell operating companies. But, in return, it was now free to compete in unregulated areas such as data processing.

As of January 1, 1984, all these operating companies throughout the country were aggregated into seven independent regional holding companies, commonly referred to in the press as the *Baby Bells*. While AT&T was no longer a monopoly company, it remained a dominant long-distance carrier, and therefore continued to be regulated. The regional holding companies constituted monopolies in their geographical areas and were also regulated. These companies, as well as the other smaller local companies, were also required to provide the same quality of connection to all the long-distance competitors. This *equal access provision* meant that you can choose your long-distance company among all the competitors with equal facility. The holding companies were themselves forbidden to compete in the long-distance market or to manufacture their own equipment. Finally, while the Bell Laboratories were allowed to remain with the reconstituted AT&T, significant segments were spun off to form a new organization called Bell Communications Research (Bellcore), to provide support to the regional holding companies.

Needless to say, the Modified Final Judgment is not the final word. Both AT&T and the Baby Bells are continually pressing for relaxation of the restrictions imposed upon them, and their competitors are equally aggressive in striving to tighten them.

Was divestiture good or bad? The answer to this question depends upon many things. Divestiture was done in the name of competition. There is certainly more of that now that the divestiture is several years old. Long-distance costs to consumers have come down significantly, in part, because of competition but also because long-distance profits can no longer subsidize local service as they once did. Thus, your phone bill for basic local service is substantially higher than it was before divestiture. Does competition necessarily mean better service for the consumer? It is clearly more complicated for the subscriber. Instead of dealing with a single entity, the subscriber now must deal with one company for local service and one of several long-distance companies. The Department of Defense opposed divestiture, because AT&T always provided excellent service in matters of telecommunications in the national interest, some of it without a direct cost to the government. It remains to be seen if a combination of companies can do as well.

While it is dangerous to draw conclusions about trends until enough time has elapsed for trends to become evident, it is nevertheless fair to say that there is no indication that AT&T or any of its communications competitors is presenting anything approaching a major challenge to the computer manufacturers in their traditional business areas. But there is also no indication that the computer companies are about to present a challenge to the carriers. IBM made an abortive attempt with a satellite-based carrier in the late 1970s, and since then has shied away from communications. Thus the fusion of the two fields, while of great technological significance, has yet to manifest itself in an industrial realignment. But it can take years for trends of such magnitude to develop, and judgment must be reserved.

In the long run the most important result of divestiture may well be the change in character of the Bell Laboratories. The AT&T Bell Laboratories—the only use of the name *Bell* allowed to AT&T by the Modified Final Judgment—have been a major national resource far exceeding their specific role as the research and development arm of the old Bell System. While their purpose was to perform the research and development necessary to sustain and advance the telephone system, the enlightened management of AT&T recognized that fundamental science was a key component of this research. The economic climate for basic research at the Laboratories in today's competitive environment is far different than before divestiture where, with the approval of the FCC, research and development expenditures were just one other set of expenses used to develop the telephone tariffs. In fact, this de facto subsidy has prompted the semifacetious observation that the Bell Laboratories were the nation's best "government" laboratory. This change may have a long-lasting effect not only on telecommunications and computing but upon our whole society as well.

The Future

When asked to make short term predictions, say five years or so, people usually overestimate the change that will take place. They tend to extrapolate the present optimistically into the future, without giving due regard to the practical impediments to change that always crop up. But in the long term the opposite is almost invariably true. Most of us are smart enough to extrapolate, but not to predict inventions or breakthroughs of other sorts that can introduce dramatic changes. We therefore have a strong tendency to underestimate the changes that will occur over, say, a 20 year span. With this in mind, what can we say about the future?

It is easy to predict that the digitization of the telecommunications networks in the form of the ISDN, or, more generally, the trend toward the fusion of communications and computing, will continue unabated. This will have significant economic consequences. It will also have philosophical and social consequences. All forms of information will be treated as computer data and transported by what appears to the consumer to be a single computer network. But it could be more than that. This single network will be able to provide us with information in addition to simple connectivity. It does not take too much imagination to note that the computers now used in the networks to improve service can also serve as information sources. Thus, the digital networks could become true information utilities—another step in the fusion of computers and communications.

How soon will all this happen? In keeping with my warning about overoptimism in the short run, it will probably take longer than the telecommunications carriers are telling us. ISDN is now available on a test basis in a limited number of locations here and abroad. Its availability will surely increase during the next two decades, but I don't believe that anyone can predict with accuracy when it will become universal. Information services are now being furnished on a small scale by several competing companies, but only minimally by the carriers.

Where else will this all lead? As significant as ISDN and the emerging information utility services may be, they are not likely to be the only new directions that we shall see in the computing/communications fusion. One strong possibility is that ultimately we will see digital television of extraordinary quality distributed to our homes by a fiber-optic cable distribution system. But I am confident that more than this will happen. When we are peering a long way into the future, who can say? New products will be marketed, and new industries will be created all directly resulting from the synergy between computers and communications.

Index